Basic English
for Physics

MAKOTO IMURA

CENGAGE
Learning™

Basic English for Physics [Text only]

Makoto Imura

© 2011 Cengage Learning K.K.

Cover Image:
© Jose Luis Pelaez/Photodisc/Getty

Photo credits appear on page 6, which constitutes a continuation of the copyright page.

For permission to use material from this textbook or product, e-mail to **eltjapan@cengage.com**

ISBN: 978-4-86312-414-1

Cengage Learning K.K.
No 2 Funato Building 5th Floor
1-11-11 Kudankita, Chiyoda-ku
Tokyo 102-0073
Japan

Tel: 03-3511-4392
Fax: 03-3511-4391

は じ め に

　理工系分野における英語の重要性は、ますます高まっています。めまぐるしい科学技術の進歩をいち早く世界に伝えるためには、国際的共通語である英語を用いることが不可欠だからです。しかし、非母語話者である私たちが、英語で、しかも専門的な内容を発表するのは容易なことではありません。ただ逆に、少し難しく考えすぎている面もあるのではないでしょうか。プレゼンテーションをするのに高度な文法知識は不要です。日常の会話では、母語話者でもほとんど10語以内の短い単純なセンテンスを用いて話しています。プレゼンテーションにおいても、原稿に頼らない口頭発表では、発話の大部分は短い単純なセンテンスからできているのです。

　この教科書（物理編）は「中学校で学んだ内容を英語で説明できる力があれば、国際的に十分通用する」という考え方のもとに、英語で発表する能力、すなわち「発信力」を養成することに主眼をおいて作成しました。本書を活用して、授業でいろいろな言語活動を行いながら、使える語彙や表現を増やし、まとまった内容を英語で説明できる力をぜひ身につけてください。

本書の構成と使い方

　本書には、10 のテーマ・ユニット、2 つのエクササイズ・ユニット、2 つの付録ユニットがあります。各テーマ・ユニットは、(1) Reading、(2) Experiment、(3) Presentation の 3 つのセクションから構成されています。

Reading

　Unit 9 と 10 を除き、基本的には中学校の理科の教科書で取り扱われている内容を、簡明な英語でまとめたリーディング・テキストです。大部分がすでに習って知っている事柄ですから、日本語に訳すことを目的とするのではなく、英語に意識を集中して読むことがポイントです。最終的には、英語で発表する能力を身につけることが目的ですので、ダウンロード音声を利用して、どんどん音読練習をするようにしてください。

▶ Vocabulary

　各ユニットのテーマにかかわる理科用語を中心に、重要語彙を日→英の順に並べてあります。日本語を見たらすぐに英語で言えるまで、練習しましょう。

▶ Useful Expressions

　本文の中にある重要な定型表現を抜き出したものです。ダウンロード音声を使って、ディクテーションをしながら、使えるようになるまで練習しましょう。

Experiment

　各ユニットのテーマにかかわる実験を取り上げ、目的（purpose）、仮説（hypothesis）、実験器具と材料（apparatus and materials）、手順（procedure）、結果（result）、結論（conclusion）の順に説明したものです。

▶ Functional Expressions
　実験手順や、装置の扱い方などを説明するときに用いられる機能的な表現をまとめたものです。ダウンロード音声を使って、ディクテーションをしながら、使えるようになるまで練習しましょう。

Presentation

　各ユニットのテーマにかかわる現象、実験、製品などについて、簡単なプレゼンテーションのサンプルを載せてあります。プレゼンテーションの方法については巻末の Appendix 1: Presentation Aid を参考にしてください。

▶ Slide
　テーマの画像を中心とするプレゼンテーション・スライドの一例です。

▶ Prompter Card
　発表の際に用いる、プロンプター・カードの一例です。プロンプター・カードを見ながら、ダウンロード音声を聞いて反復練習やシャドーイングを行い、最終的にはプロンプター・カードを見ながら、自分で発表ができるようにしましょう。

▶ Q & A
　質疑応答の例が日本語で書かれています。自分でまず英語に直したのち、ダウンロード音声で確認してみましょう。

［コラム］
　各ユニットの最後には、人物伝を中心としたコラムが載せてあります。真理の探究に情熱をかけた科学者たちのエピソードを読んでみましょう。

［テーマ・ユニット以外のユニット］
Exercises 1, 2
　5 ユニットごとに、まとめの練習問題を配置してあります。理解の確認とともに、質疑応答の練習用としても活用してください。

Appendix 1: Presentation Aids
　プレゼンテーションの方法について解説してあります。グループ発表などの学習活動を行う際に、参考にしてください。

Appendix 2: Glossary
　理科用語集です。用例とともに覚えるようにしてください。

Contents

What Is Physics?

重力

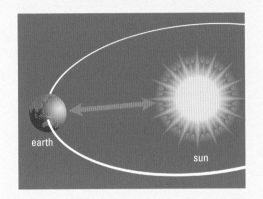

gravity

電磁気力

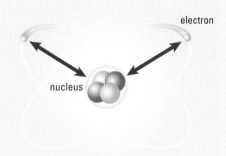

electron

nucleus

electromagnetism

弱い力

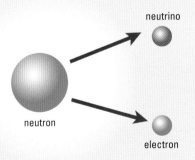

neutrino

neutron

electron

weak interaction

強い力

nucleus
(protons, neutrons)

strong interaction

CONTENTS

1. Classical Physics and Modern Physics
2. In This Textbook

Reading

Classical Physics and Modern Physics

Physics is the scientific study of matter and energy. It deals with all aspects of nature from the structure of the atom to the workings of the universe. The scientific revolution that took place during the **Renaissance** transformed the fundamental assumptions about the universe. Copernicus (1473–1543) questioned the **geocentric model** (Ptolemaic system) of the

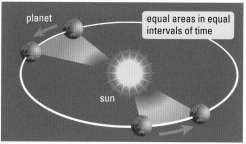

Fig. 1 Kepler's Second Law

solar system and advocated the **heliocentric model** (Copernican system). Kepler (1571–1630) discovered the laws governing planetary revolutions around the sun (Fig. 1). Galileo (1564–1642) demonstrated the truth of the Copernican system by observing **celestial bodies** using his own telescope. Galileo also conducted numerous experiments on **mechanics** including those on **free fall** and **uniform**

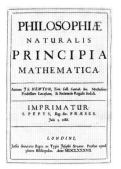

Fig. 2 Principia

linear motion. Newton (1642–1727) combined the works of these forerunners and established the foundation of **classical physics** (Fig. 2)[1]. Classical physics deals with subjects that had been well developed by the 1900s, which include mechanics, heat, sound, light, electricity, and magnetism. By then, it seemed that there remained little to be discovered in physics. At the beginning of the 1900s, however, scientists found things that could not be explained by classical physics, which gave birth to **modern physics**. The epoch-making discovery of the **nucleus** by Rutherford (1871–1937) in 1911 led to the study of

elementary particles in **nuclear physics**. Then Einstein's (1879–1955) **theory of relativity**, along with a series of laws about energy, mass, space, time, and the speed of light, changed almost everything that scientists had believed or understood before. Together with Planck (1858–1947), Einstein also contributed to the development of **quantum theory**, which originated from the question of whether light is a wave or a stream of particles.

(The double-slit experiment in Fig. 3 demonstrates that light has characteristics of both waves and particles.)[2] Quantum theory is a complicated theory, but it sheds new light on the ideas of matter and energy, gradually unlocking the deepest secrets of the universe.

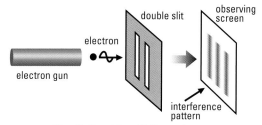

Fig. 3 Double-slit Experiment

In This Textbook

Apart from the last unit, this textbook mainly deals with the subjects of classical physics. It begins with the study of light (Unit 2), which has also played a vital role in the development of modern physics. Unit 3 features sound, which tells us the characteristics of wave motion. Units 4 and 5 explain force and motion, which constitute the heart of mechanics. Units 6 and 7 introduce electricity and electromagnetism. Units 8 and 9 are about work and energy, with particular emphasis on the study of heat (thermodynamics) in Unit 9. The final unit (Unit 10) gives a brief introduction of modern physics.

NOTES

1) 写真は『プリンキピア』として知られるニュートンの著作（1687）。この中でニュートンは、万有引力の法則（the law of universal gravitation）を発表した。

2) 光の代わりに、粒子であることが分かっている電子を、電子銃を使って1個ずつ飛ばして観察する実験。二重スリットの後ろにあるスクリーンには、波の特長であるしま模様が現れる。

Vocabulary

☐ ルネサンス	Renaissance	☐ 等速直線運動	uniform linear motion
☐ 天動説	geocentric model (Ptolemaic system)	☐ 古典物理学	classical physics
		☐ 現代物理学	modern physics
☐ 地動説	heliocentric model (Copernican system)	☐ 原子核	nucleus
		☐ 素粒子	elementary particle
☐ 天体	celestial body	☐ 原子物理学	nuclear physics
☐ 力学	mechanics	☐ 相対性理論	theory of relativity
☐ 自由落下	free fall	☐ 量子論	quantum theory

Useful Expressions

1) Physics (   ) (   ) all aspects of nature.（物理学は、自然のすべての側面を取り扱う）

2) The discovery of the nucleus (   ) (   ) the study of elementary particles.（原子核の発見が、素粒子の研究につながった）

3) Quantum theory (   ) (   ) (   ) (   ) the ideas of matter and energy.（量子論は、物質とエネルギーに対する考え方に新たな光明を投ずるものである）

4) Light has (   ) (   ) vital (   ) in the development of modern physics.（光は、現代物理学の発展に重要な役割を果たした）

物理学の先人たち

Thales

Pythagoras

Heraclitus

Democritus

Aristotle

物理学の源流は、万物の根源を探究した古代ギリシャの自然哲学にさかのぼる。Physics はギリシャ語で「自然」を意味する *physis*（*φύσις*）に由来している。「最初の哲学者」と呼ばれるタレス（Thales c. 624 BC–c. 545 BC）は、万物の根源（アルケー）を水であると考え、ヘラクレイトス（Heraclitus c. 540 BC–c. 480 BC）は、それを火であると考えた。また、三平方の定理（the Pythagorean theorem）で有名なピタゴラス（Pythagoras c. 580 BC–c. 500 BC）は、アルケーを数であると考えた。例えば、彼は音の研究で協和音が振動数の整数倍になっていること（ピタゴラス音律）を発見し、自然が数の調和によって成り立っていると確信していた。驚くべきことに、原子論はすでにこの時代に誕生している。デモクリトス（Democritus c. 460 BC–c. 370 BC）は、物質がこれ以上分けることのできない粒子である原子（atom）から成るとした。「学問の祖」と称されるアリストテレス（Aristotle 384 BC–322 BC）は、四元素説を唱え、地上の物は、土・水・空気・火の4つの元素の混合によってできていると考えた。彼はさらに、宇宙には第5の元素であるエーテル（ether）が満ちていると考えた。エーテルは光の波を伝えるために必要な媒質として、19世紀末までその存在が信じられていた物質である。

Unit 2

Light

反射

reflection

屈折

refraction

スペクトル

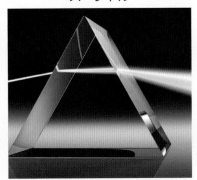

spectrum

光ファイバー

optical fiber

CONTENTS

1. Characteristics of Light
2. Lenses and Images
3. Color

◉Reading

Characteristics of Light

We can see things because of light. When light hits an object, it is scattered in every direction (**diffuse reflection**). The diffused light goes through the lenses of our eyes and then forms an image of the object on the **retina**. When a beam of light strikes a polished surface such as a mirror at a certain angle, it is reflected at the same angle. Figure 1 shows how light is reflected by a mirror. The angle of the incoming light against the **normal** (the line perpendicular to the surface at the point of **incidence**) is called the **angle of incidence**, and the angle of reflected light against the normal is called the **angle of reflection**. Light travels straight, but it can bend when it goes through different substances. This phenomenon is called **refraction**. Refraction occurs because the speed of light[1] changes according to the density of the **medium**. Figure 2 shows how light is refracted between air and water. The angle of refracted light against the normal is called the **angle of refraction**. When the angle of incidence goes beyond a certain angle, the light no longer passes through the surface but is completely reflected. This phenomenon is called **total reflection**.

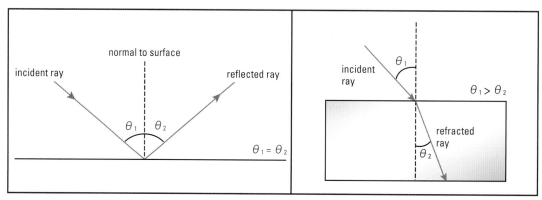

Fig. 1 Reflection   Fig. 2 Refraction

The angle of incidence is the same as the angle of reflection (Fig. 1). The angle of incidence is greater than the angle of refraction when light passes from a low-density medium (e.g. air) to a high-density medium (e.g. water) (Fig. 2).

Lenses and Images

There are two types of lenses: **convex lenses** and **concave lenses**. A convex lens is thicker in the middle while a concave lens is thinner in the middle. As Fig. 3 shows, all the parallel rays of light passing through a convex lens come to a focal point called the **focus**. If an object is placed outside the focus, an inverted image of the object is formed on the other side of the lens. This image is called a **real image**

(Fig. 3-a). On the other hand, if an object is placed between the focus and the lens, an upright magnified image is formed on the same side as the object. This image is called a **virtual image**, because it is an illusionary image created by refraction (Fig. 3-b).

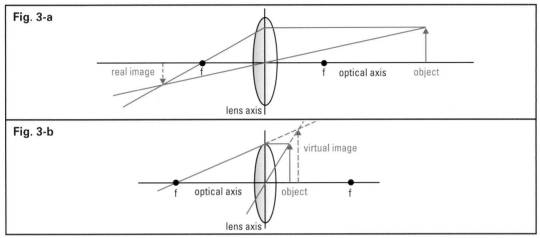

Fig. 3 Convex Lens and Real Image / Virtual Image

Color

Sunlight or white light is composed of gradations of light with varying **wavelengths**, which we perceive as colors. When a beam of sunlight is sent through a prism, it is broken up into different colors because light is bent differently according to wavelength. This band of colors is called a **spectrum** (Fig. 4)[2]. The red light, which has the longest wavelength and thus has the lowest **frequency**, is bent the least, while the violet light, which has the shortest wavelength and thus has the highest frequency, is bent the most. Actually, there is an extended portion on each side of the visible spectrum, the **infrared** and the **ultraviolet**, which are not visible to human eyes. We perceive something as being red because the material of the object reflects only the red light but absorbs all the other colors.

Fig. 4 Spectrum

NOTES

1) 光の速さは、真空中で秒速約30万kmである。
2) 英語圏の子どもたちは、次の文を暗記して虹の色を覚える。Richard Of York Gave Battle In Vain. (R = Red, O = Orange, Y = Yellow, G = Green, B = Blue, I = Indigo, V = Violet.)

Vocabulary

☐ 乱反射	diffuse reflection	☐ 凸レンズ	convex lens
☐ 網膜	retina	☐ 凹レンズ	concave lens
☐ 法線	normal	☐ 焦点	focus
☐ 入射	incidence	☐ 実像	real image
☐ 入射角	angle of incidence	☐ 虚像	virtual image
☐ 反射角	angle of reflection	☐ 波長	wavelength
☐ 屈折	refraction	☐ スペクトル	spectrum
☐ 媒質	medium	☐ 周波数	frequency
☐ 屈折角	angle of refraction	☐ 赤外線	infrared
☐ 全反射	total reflection	☐ 紫外線	ultraviolet

Useful Expressions

1) A real image of an object is formed (　　　　) (　　　　) (　　　　)
   (　　　　) (　　　　) the lens. （物体の実像は、レンズの<u>反対側に</u>生じる）

2) A virtual image is formed (　　　　) (　　　　) (　　　　) (　　　　)
   (　　　　) the object. （虚像は、物体<u>と同じ側に</u>生じる）

3) White light (　　　　) (　　　　) (　　　　) gradations of light with
   varying wavelengths. （白色光は、波長の異なった光のグラデーション<u>によって</u>
   構成されている）

◆Experiment

☐ Purpose: To observe how light is refracted.
☐ Apparatus and materials: a ten-yen coin, a coffee cup, water
☐ Procedure:
- Place a ten-yen coin at the bottom of a coffee cup.
- Look at the cup from an angle, and see how the coin comes into view
  as your partner pours water into the cup.

Refraction Experiment

☐ Result: At first the coin is not visible, but as water is poured into the cup,
the figure of the coin appears gradually. This occurs because the light is
bent when it comes out of water. What we see is a virtual image of the coin.

Functional Expressions

1) Observe (   ) light (   ) (   ). （光がどのように屈折するか
   観察する）

2) (   ) a ten-yen coin (   ) the bottom of a coffee cup. （10 円玉
   をコーヒーカップの底に置く）

3) (   ) water (   ) the cup. （カップに水を注ぐ）

◉Presentation

Slide

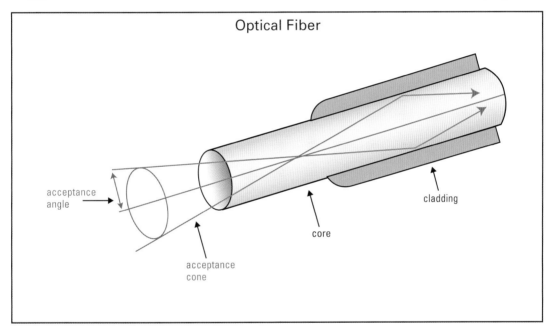

Prompter Card

- An optical fiber is a device to transmit light signals through a glass or
  plastic fiber.
- It consists of the core and the cladding.
- Light is kept inside the core by total internal reflection.
- An optical fiber can transmit a larger amount of data over longer distances.

15

Q & A

Q. なぜ光ファイバーは、多くのデータを、遠くまで運べるのですか。

A. 光ファイバーで用いる光波は、波長が約 1.3 〜 1.5 ミクロンの超高周波です。したがって、波長がより長い電波に比べて、単位時間内に多くのデータを送ることができます。また全反射の反射率は 100% なので、光の信号は遠くまで弱まることなく伝わります。

学習活動

1) 光に関する現象、実験、製品などについてワンポイント・スライドを作成してみよう。

2) プロンプターカードを作成して、ミニ・プレゼンテーションをしてみよう。

3) 英語で質問や答えを作成して、互いに質疑応答の練習をしてみよう。

光の正体

光は粒子（particle）なのか、波動（wave motion）なのかという問題は、物理学上の大きな謎である。かつてニュートン（Isaac Newton 1642–1727）は粒子説を唱え、ホイヘンス（Christiaan Huygens 1629–1695）は波動説を唱えた。現代では、光は電磁波（electromagnetic waves）の一種であることが分かっているが、同時に光の粒子性を考慮しなければ説明のつかないことも多い。

Newton

Huygens

アインシュタイン（Albert Einstein 1879–1955）はプランク（Max Planck 1858–1947）のエネルギー量子仮説（energy quantum hypothesis）をもとに、光を、エネルギーを持つ粒子の流れであると考えた（光量子仮説：light quantum hypothesis）。この粒子のことを光子（photon）と言う。光の性質の二重性を、ミクロの世界の本質的な現象として解明するために、量子力学（quantum mechanics）という学問が生まれることになった。

同じ波動でも、音波は縦波（longitudinal wave）で、空気や水のような媒質（medium）がなければ伝わらないのに対して、光は横波（transverse wave）で、媒質のない真空中でも伝わる。また、光はエネルギーを持っており、私たちは太陽光を熱や電気に変換して利用している。光の速さは秒速 30 万 km で、1 秒で地球を 7 回半回り、月まで（約 38 万 km）およそ 1.3 秒で到達し、太陽の光が地球に届くまで約 8 分半かかる。光の速度は一定で、また光より速いものは存在しないことが分かっており、このことからアインシュタインは、相対性理論（theory of relativity）を打ち立てることになる。

Planck

Einstein

Unit 3

Sound

音波

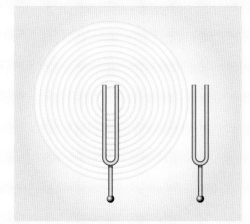

sound waves

波長と振幅

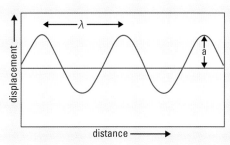

λ = wavelength
a = amplitude

wavelength / amplitude

オシロスコープ

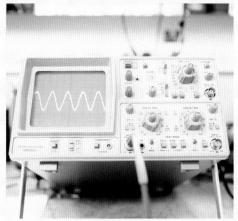

oscilloscope

超音波

ultrasound

CONTENTS

ⓘ Reading

Characteristics of Sound

Sound is caused by **vibration** (oscillation). We hear the sound of a bell because the bell's vibration is **transmitted** through the air and reaches our **eardrums**. Sound also travels through other substances such as water and metals, but it cannot travel in a vacuum because there is no **medium** for transmitting the vibration. The speed of sound is approximately 340 m/s (1,200 km/h) through the air, and it travels four times faster through water and 15 times faster through iron. Figure 1 shows how the vibration of a **sound fork** is transmitted through the air. The vibrating sound fork pushes and pulls the air around it causing a series of condensed and rarefied areas of air. The waves formed in this way are known as **compression waves**.

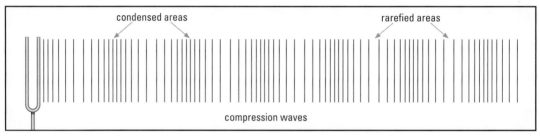

Fig. 1 Sound Waves

Sound Waves and Oscilloscope

Sound waves are **longitudinal waves**, which means the direction of vibration is parallel to the direction of transmission. To make observation easier, an **oscilloscope** translates sound waves and displays them as if they were **transverse waves** in which the direction of vibration is perpendicular to the direction of transmission (Fig. 2). The length of a single wave is called the **wavelength** (λ), which is equal to the distance between two adjacent peaks (**crests**) or valleys (**troughs**). The number of vibrations per second is called the **frequency**. **Hertz** (Hz) is a unit of frequency and 1 Hz means one wave cycle (= λ) per second. The height of a crest or the depth of a trough is called the **amplitude**.

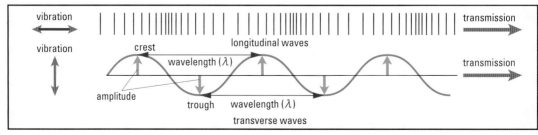

Fig. 2 Longitudinal Waves / Transverse Waves

Three Elements of Sound

The three elements of sound are **pitch**, **loudness**, and **tone**. (1) Pitch is determined by the frequency of sound waves. The higher the frequency is, the higher the pitch becomes, and the lower the frequency, the lower the pitch. Humans can hear sound with frequencies between 20 Hz and 20,000 Hz. Sound lower than 20 Hz is called **infrasound**, while sound higher than 20,000 Hz is called **ultrasound**. Animals like bats can use ultrasound to detect objects and navigate their way through the darkness. (2) Loudness is determined by the amplitude of sound waves. The greater the amplitude is, the louder the sound becomes. Loudness, however, is perceived rather subjectively. Our ears tend to perceive a tenfold increase in sound as only about a doubling or tripling in sound. The **decibel** (dB) is used as a unit to represent the magnitude of sound in a way closer to how humans perceive the loudness of sound. It is a **logarithmic** ratio and a difference of 10 dB means an intensity of 10 times. Therefore, a sound of 20 dB is 10 times more intense than a sound of 10 dB although it may be perceived to be only about twice as loud by human ears. Our everyday conversation is about 60 dB, while industrial noise can reach 90 dB, which is actually 1,000 times more intense than conversation. (3) Tone is determined by the shape of sound waves. Figure 3 shows the sound waves of various kinds of musical instruments.

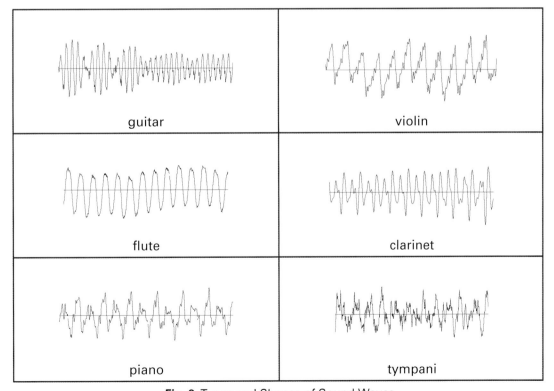

Fig. 3 Tones and Shapes of Sound Waves

Vocabulary

☐ 振動	vibration (oscillation)		☐ （波の）谷	trough
☐ 伝わる	transmit		☐ 周波数	frequency
☐ 鼓膜	eardrum		☐ ヘルツ	hertz (Hz)
☐ 媒質	medium (*pl.* media)		☐ 振幅	amplitude
☐ 音叉	sound fork		☐ 音の高さ	pitch
☐ 疎密波	compression wave		☐ 音の強さ	loudness
☐ 縦波	longitudinal wave		☐ 音色	tone
☐ オシロスコープ	oscilloscope		☐ 超低周波音	infrasound
☐ 横波	transverse wave		☐ 超音波	ultrasound
☐ 波長	wavelength		☐ デシベル	decibel (dB)
☐ （波の）山	crest		☐ 対数の	logarithmic

Useful Expressions

1) Sound cannot travel (   ) (   ) (   ).（音は真空中を伝わらない）

2) In longitudinal waves, the direction of vibration is (   ) (   ) the direction of transmission.（縦波では、振動の方向は波が伝わる方向に対して平行である）

3) In transverse waves, the direction of vibration is (   ) (   ) the direction of transmission.（横波では、振動の方向は波が伝わる方向に対して垂直である）

⊕Experiment

☐ Purpose: To observe how sound waves are transmitted through water.

☐ Hypothesis: Water moves along in the direction of sound wave transmission.

☐ Apparatus and materials: a plastic bowl, a polystyrene pellet, a sound fork, water

☐ Procedure:
- Fill a plastic bowl with water and float a small polystyrene pellet.
- Place a vibrating sound fork in the water and cause ripples.
- Observe how the ripples move and how the pellet moves.

Sound Wave Experiment

☐ Result: Ripples spread in a radial fashion. The pellet moves up and down, but it does not move in the direction of the waves.

☐ Conclusion: The hypothesis is rejected. Although vibrations are transmitted through a medium, the medium itself will not move away as a result of sound wave transmission.

Functional Expressions

1) (　　　　　) a plastic bowl (　　　　　) water.（プラスティックのボウルを<u>水で満たす</u>）

2) The pellet moves (　　　　) and (　　　　). （小球は<u>上下に動く</u>）

3) The hypothesis is (　　　　). （仮説は<u>棄却される</u>）

◉ Presentation

Slide

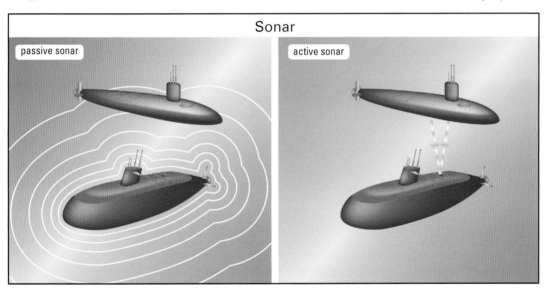

Prompter Card

- Sonar is a technology that uses sound waves to detect objects under water.
- It is also used for underwater communication.
- There are two kinds of sonar: active and passive.
- Active sonar emits pulses of sound waves and listens for the echoes.
- Passive sonar listens for sounds without emitting sound waves.

Q & A

Q. ソナーでどうやって、目的物までの距離を測るのですか。

A. ソナーの原理は、やまびこと同じです。アクティブ・ソナーでは音を発してから
その音が反射して戻ってくるまでの時間を計り、その時間に音速を掛けることに
よって往復の距離を計算することができます。水中における音速は約 1,500 m/s な
ので、時間差が 2 秒であれば、その物体は約 1,500 m 離れていることになります。

学習活動

1) 音に関する現象、実験、製品などについてワンポイント・スライドを作成してみ
よう。

2) プロンプターカードを作成して、ミニ・プレゼンテーションをしてみよう。

3) 英語で質問や答えを作成して、互いに質疑応答の練習をしてみよう。

音の探究者たち

Pythagoras

Mersenne

Bell

音の研究は古代ギリシャにさかのぼる。三平方の
定理（Pythagorean theorem）で有名なピタゴラス
（Pythagoras c. 580 BC–c. 500 BC）は、弦楽器が心地
良い協和音を奏でるとき、弦の長さが単純な整数比
になっていることを示した。イタリアの天文学者ガ
リレオ・ガリレイ（Galileo Galilei 1564–1642）は、音
の高さが弦の振動数で決まることを見いだし、フラ
ンスの数学者・物理学者メルセンヌ（Marin Mersenne
1588–1648）は、楽器の音色が、基本音とその整数
倍の振動数の音が合成されたものであることを示し
た。ドップラー効果（Doppler Effect）は、オースト
リアの物理学者クリスチャン・ドップラー（Johann
Christian Doppler 1803–1853）によって発見された。
音の大きさの単位であるデシベルは、電話機の発明で
有名なイギリスのグラハム・ベル（Alexander Graham
Bell 1847–1922）に、周波数の単位であるヘルツは、
ドイツの物理学者ハインリッヒ・ヘルツ（Heinrich
Rudolf Hertz 1857–1894）にそれぞれちなんだもので
ある。

Galilei

Doppler

Hertz

Force

力の３要素

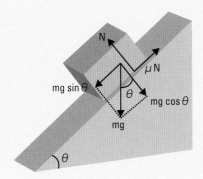

three elements of force

振り子

pendulum

力のつり合い

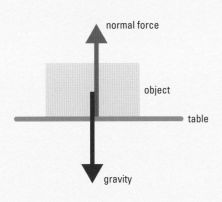

equilibrium

作用と反作用

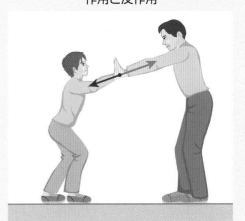

action and reaction

CONTENTS

⊙ Reading

Forces around Us

A force is the ability to produce physical changes; it acts on an object and **deforms** or **supports** it, or changes its motion. There are two types of forces, ones that act by direct contact (**direct forces**) and ones that act at a distance without direct contact (**forces at a distance**). The direct forces include **pushing force, pulling force, colliding force, air resistance, friction, normal force**, etc. The forces at a distance are **gravity, electromagnetism**, the **strong interaction (nuclear force)**, and the **weak interaction**. The four forces at a distance are also called the **fundamental forces**. Things fall to the ground because the earth attracts objects. This attractive force, known as gravity, is proportional to the **mass** of the object. Therefore, the force of gravity acting on an object of 2 kg is twice as large as the force of gravity acting on an object of 1 kg. The SI unit[1] of force is the **newton**[2] (N), and 1 N is about the same as the gravitational force[3] acting on an object of 100 g.

Three Elements of Force

A force can be represented by an arrow. The back-end of the arrow indicates the **point of application of force**, the arrowhead indicates the **direction of force**, and the length of the arrow represents the **magnitude of force** (Fig. 1).

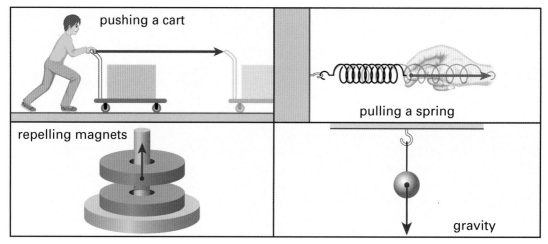

Fig. 1 Force Arrows

When two forces acting on an object are equal in magnitude but opposite in direction, the forces are balanced and they cancel out each other. This state is called **equilibrium**. An object at equilibrium is either static or moving at a constant speed (Fig. 2). Now, equilibrium is often confused with **action and reaction**, but they are completely different concepts. While equilibrium applies to a single object, action and reaction applies to two objects interacting with each other. Whenever one object exerts a force on the other (action), the second object exerts an equal and opposite force on the first (reaction). In other words, forces always act in pairs, and for every action, there is always an opposite and equal reaction. Since the action and reaction forces work on different bodies, they do not cancel out each other (Fig. 3).

Fig. 2 Equilibrium of Forces

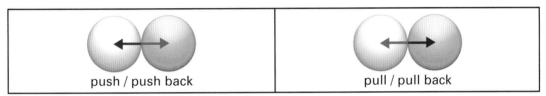

Fig. 3 Action and Reaction

Pressure

Pressure is the amount of force applied per unit area (Fig. 4-a). For example, if 1 N of force is applied to the area of 1 m², the pressure is 1 N/m². The SI unit of pressure is the **pascal** (Pa), and 1 Pa equals 1 N/m². **Atmospheric pressure** is often measured in **hectopascals** (hPa). One hectopascal (1 hPa) is 100 Pa. The atmospheric pressure at sea level is 1,013 hPa, which is about the same as the pressure of 1 kg of weight per cm² (Fig. 4-b).

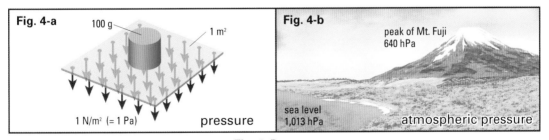

Fig. 4 Pressure

NOTES

1) 国際単位（International System of Units）
2) 1 N は、質量 1 kg の物体に 1 m/s² の加速度を与える力である。（1 N = 1 kg·m/s²）
3) 地球の重力の大きさは、9.8 N/kg である。したがって、100 g の物体には約 1 N の重力がかかっていることになる。

Vocabulary

☐	…を変形する	deform	☐ 弱い相互作用	weak interaction
☐	…を支える	support	☐ 基本相互作用	fundamental forces
☐	接触力	direct forces	☐ 質量	mass
☐	場の力	forces at a distance	☐ ニュートン	newton (N)
☐	押す力	pushing force	☐ 力の作用点	point of application
☐	引く力	pulling force		of force
☐	衝突力	colliding force	☐ 力の向き	direction of force
☐	空気抵抗	air resistance	☐ 力の大きさ	magnitude of force
☐	摩擦力	friction	☐ つり合い・平衡	equilibrium
☐	垂直抗力	normal force	☐ 作用と反作用	action and reaction
☐	重力	gravity	☐ 圧力	pressure
☐	電磁気力	electromagnetism	☐ パスカル	pascal (Pa)
☐	強い相互作用	strong interaction	☐ 大気圧	atmospheric pressure
☐	核力	nuclear force	☐ ヘクトパスカル	hectopascal (hPa)

Useful Expressions

1) The gravitational force is (   ) (   ) the mass of objects. （物体に働く重力の大きさは、その物体の質量に比例する）

2) Equilibrium occurs when two forces acting on an object are equal (   ) magnitude but opposite (   ) direction.（力のつり合いは、1つの物体に働く力の大きさが等しく、方向が反対の場合に起こる）

3) The normal atmospheric pressure at sea level is about the same as the pressure of 1 kg of weight (   ) cm^2.（海抜高度ゼロでの通常の気圧は、およそ 1 cm^2 あたり 1 kg の圧力に等しい）

⊕Experiment

☐ Purpose: To measure action and reaction forces between two objects.

☐ Hypothesis: When two objects collide with each other, the heavier object will exert a greater force.

☐ Apparatus and materials: two force sensors, two toy carts, weights

☐ Procedure:

• Prepare two toy carts and load one of them with weights.

• Let them collide head-on into each other.

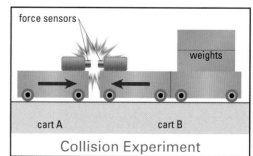

Collision Experiment

- Measure the force exerted on each cart using force sensors.
☐ Result: The forces exerted on both carts are exactly the same.
☐ Conclusion: The hypothesis is rejected. Action and reaction are always equal in strength and opposite in direction.

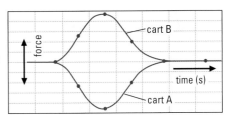

Functional Expressions

CD-27

1) (   ) the cart (   ) weights.（台車に<u>重り</u>を<u>積む</u>）

2) Let them (   ) (   ) into each other.（それらを<u>正面衝突</u>させる）

3) The two forces are equal (   ) (   ).（2つの力は<u>強さ</u>において等しい）

◉ Presentation

Slide

CD-28

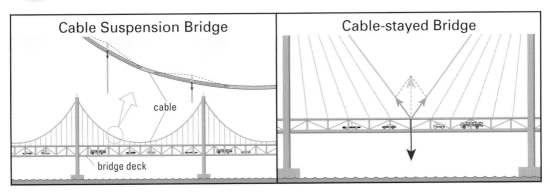

Cable Suspension Bridge — cable, bridge deck

Cable-stayed Bridge

Prompter Card

- There are two types of cable bridges: cable suspension bridge and cable-stayed bridge.
- How do these bridges support the bridge deck with cables?
- The weight of the bridge deck is divided into two component forces.
- The resultant force is represented by the diagonal vector of a parallelogram.
- The vector is equal in size to the downward force and opposite in direction.

Q. なぜ瀬戸大橋の主塔は、あんなに高いのですか。

A. 主塔が低いと、ケーブルの張力が大きくなって、橋げたを支えきれなくなるからです。

学習活動

1) 力に関する現象、実験、製品などについてワンポイント・スライドを作成してみよう。

2) プロンプターカードを作成して、ミニ・プレゼンテーションをしてみよう。

3) 英語で質問や答えを作成して、互いに質疑応答の練習をしてみよう。

巨人の肩に乗る

Aristotle

Galilei

Newton

古代ギリシャ時代、アリストテレス（Aristotle 384 BC–322 BC）は、重いものほど速く落下すると考えていた。そのおよそ2000年後にこの説をくつがえしたのが、ガリレオ・ガリレイ（Galileo Galilei 1564–1642）である。ガリレオは、ピサの斜塔から大小2つの球を落として、同時に着地することを証明したといわれている。物体が質量に関係なく等加速度（constant acceleration）で自由落下（free fall）することは、すなわち質量の大きいものほど動かすのに大きな力がかかることを意味しており、これはニュートンの運動の第2法則（運動方程式 $F = ma$）につながっていく。またガリレオは、運動の第1法則（慣性の法則：the law of inertia）を予見する実験も行っている。デカルト（René Descartes 1596–1650）は運動する物体の「勢い」あるいは物体の運動を変化させるときにかかる瞬間の「衝撃」というものを考えたが、これはニュートン力学における運動量（momentum）と力積（impulse）の考え方につながる。これらに運動の第3法則（作用・反作用の法則：the law of action and reaction）を加えて、ニュートンの「運動の法則」が完成する。ニュートンは、友人のフック（Robert Hooke 1635–1703）へあてた手紙の中で、"If I can see things further than anyone else, it is only because I'm standing on the shoulders of giants." と書いている。

Descartes

Hooke

Motion

等速直線運動

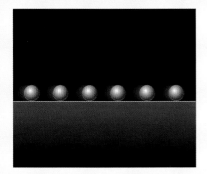

uniform linear motion

等加速度運動

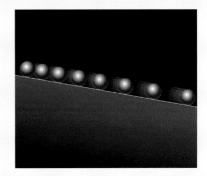

uniform accelerated motion

自由落下

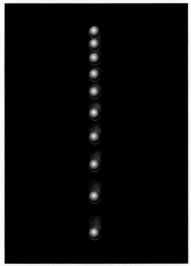

free fall

放物運動

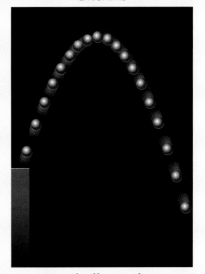

parabolic motion

Reading

Motion

Force makes things move. If a force is applied once to an object, the object will start moving and keep on moving at a constant speed if there is no friction. For example, if you throw a ball in **outer space** where there is no friction, the ball will keep on moving straight forever at a constant speed according to the **law of inertia**. This motion is called **uniform linear motion**. The **distance** traveled in uniform linear motion is calculated by multiplying **speed** and time ($d = vt$, where d is distance, v is speed or **velocity**[1], t is time) (Fig. 1-a). On the other hand, if a constant force is applied to an object, the object will start moving and keep on accelerating. This motion is called **uniform accelerated motion**. The distance traveled in uniform accelerated motion is calculated by the following equation: $d = \frac{1}{2} at^2$, where d is distance, a is **acceleration**, t is time (Fig. 1-b).

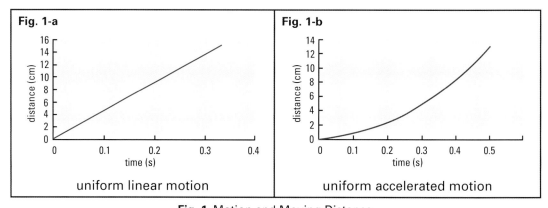

Fig. 1 Motion and Moving Distance

Force and Motion

The larger the mass of an object, the more force is needed to move it, which means force is proportional to the mass of an object ($F \propto m$). At the same time, we need more force to give more acceleration to an object, which means force is proportional to acceleration as well ($F \propto a$). From these two conditions, Isaac Newton established the **equation of motion** ($F = ma$), and he determined the newton (N) as a unit of force, where 1 N is the force which gives a 1 kg mass an acceleration of 1 m/s² (one meter per second per second). For example, in order to give an acceleration of 10 m/s² to an object with 10 kg, we need a force of 100 N (10 kg × 10 m/s²). According to the equation of motion, no force means no acceleration. That is to say, unless any outside force is applied, an object either stays at rest or keeps on moving at a constant speed. These two states are physically the same. They are both in a state of equilibrium and the law of inertia applies (Fig. 2).

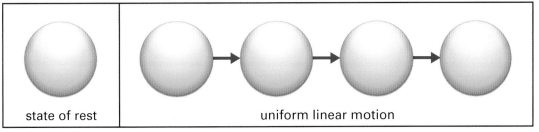

state of rest | uniform linear motion

Fig. 2 State of Equilibrium

Gravity and Free Fall

Gravity is the force of attraction between all masses in the universe, but we often refer to it as the earth's pull on things. Gravity causes all objects to accelerate to 9.8 m/s². It means the effect gravity has on an object of 1 kg equals 9.8 N. Gravity is a constant force that makes all things fall. The uniform accelerated motion due to gravity is called **free fall**. Since **gravitational acceleration** is 9.8 m/s², the **fall velocity** is given by 9.8 × time ($v = at$) and the **fall length** is given by $\frac{1}{2}$ × 9.8 × time squared ($d = \frac{1}{2} at^2$). For example, if you drop a stone from a cliff, the velocity in 5 seconds will be 49 m/s (9.8 m/s² × 5 s) or about 176 km/h, and the fall length will be 122.5 m [$\frac{1}{2}$ × 9.8 m/s² × (5 s)²] (Fig. 3-a). Then, what happens if you throw a stone **horizontally** from a cliff? There are two forces acting on the stone (excluding **air resistance**). One is a **thrust** or a one-time force that gives a uniform linear motion to the stone (Fig. 3-b), and the other is gravity or a constant force that gives a constant accelerated motion to the stone (Fig. 3-a). These two forces are combined to make the stone fall in a **parabolic motion** as in Fig. 3-c.

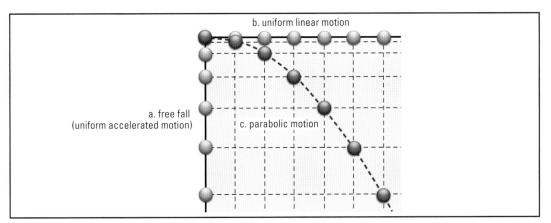

Fig. 3 Free Fall and Parabolic Motion

NOTES

1) 速さ（speed）はスカラー値（scalar quantity）で、速度（velocity）はベクトル値（vector quantity）である。つまり速さは大きさ（size）しか持たないのに対して、速度は大きさと向き（direction）を持つ。

☐ 宇宙空間	outer space	☐ 重力（引力）	gravity
☐ 慣性の法則	law of inertia	☐ 自由落下	free fall
☐ 等速直線運動	uniform linear motion	☐ 重力加速度	gravitational acceleration
☐ 距離	distance	☐ 落下速度	fall velocity
☐ 速さ	speed	☐ 落下距離	fall length
☐ 速度	velocity	☐ 水平方向に	horizontally
☐ 等加速度運動	uniform accelerated motion	☐ 空気抵抗	air resistance
		☐ 推力	thrust
☐ 加速（度）	acceleration	☐ 放物運動	parabolic motion
☐ 運動方程式	equation of motion		

Useful Expressions

1) The distance traveled in uniform linear motion is calculated by (   ) speed (   ) time. （等速直線運動における移動距離は、速さ<u>と</u>時間<u>をかけ合わせる</u>ことによって計算される）

2) Unless any outside force is applied, an object either stays (   ) (   ) or keeps on moving at a constant speed. （外部から力が加わらなければ、物体は<u>静止したままか</u>、あるいは等速で動き続ける）

3) When we say "gravity," we often (   ) (   ) (   ) (   ) the earth's pull on things. （「重力」と言うとき、私たちはしばしば地球の引力<u>のことを指して言う</u>）

⬥Experiment

☐ Purpose: To measure the acceleration of gravity.
☐ Apparatus and materials: a digital camera, a tripod, a stroboscope, a meter stick (with a stand and a weight holder), a steel ball
☐ Procedure:
 • Set up a meter stick vertically against a wall with a dark background.
 • Place a digital camera on a tripod and aim toward the meter stick.
 • Place a stroboscope behind the camera and set the strobe rate at 0.05 seconds.
 • Darken the room and photograph the fall of the steel ball at slow shutter speed.

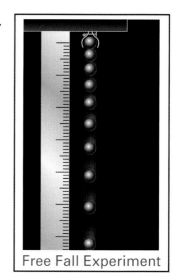

Free Fall Experiment

☐ Result: (example)

Time (s)	Location (m)
0.00	0.00
0.05	0.04
0.10	0.11
0.15	0.20
0.20	0.32
0.25	0.46
0.30	0.62
0.35	0.81

Average Speed (m/s)
(a)
1.4
1.8
2.4
2.8
3.2
(b)

Acceleration

$$\frac{(b.\quad) \, m/s - (a.\quad) \, m/s}{0.3 \, s}$$

$= (c.\quad) \, m/s^2$

☐ Conclusion: The acceleration due to gravity is approximately (c.   ) m/s².

Functional Expressions   CD-35

1) Set up a meter stick (   ) against a wall.（壁を背にして、メートル物差しを<u>垂直に設置する</u>）

2) (   ) the camera (   ) the meter stick.（カメラをメートル物差しの方へ向ける）

3) (   ) the strobe rate (   ) 0.05 seconds.（ストロボの速さを 0.05 秒に設定する）

⊙ Presentation

Slide

Prompter Card   CD-36

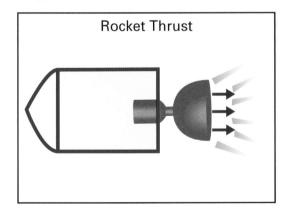

Rocket Thrust

- How can a rocket accelerate in outer space?
- Remember the law of action and reaction.
- Example: If you throw a ball in outer space, the ball pushes you back.
- A rocket pushes on exhaust gases and they push back on the rocket.

Q. ロケットで宇宙空間に噴射された燃料ガスは、その後どうなるのですか。

A. 気圧のない真空の宇宙空間で噴射された燃料ガスは、急速に拡散してしまいます。

学習活動

1) 運動に関する現象、実験、製品などについてワンポイント・スライドを作成してみよう。

2) プロンプターカードを作成して、ミニ・プレゼンテーションをしてみよう。

3) 英語で質問や答えを作成して、互いに質疑応答の練習をしてみよう。

万有引力と重力

万有引力の法則（the law of universal gravitation）は、アイザック・ニュートン（Isaac Newton）が、1687年に *Philosophiae Naturalis Principia Mathematica*（自然哲学の数学的諸原理）の中で発表したもので、質量を持つすべての物体は、互いに引き合うことを示している。互いの距離が r（m）離れている、質量 m_1（kg）と m_2（kg）の2つの物体に働く万有引力は、次の式で表される。

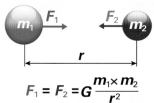

$$F_1 = F_2 = G \frac{m_1 \times m_2}{r^2}$$

すなわち、万有引力は2つの物体の質量の積に比例し、互いの距離の2乗に反比例する（G は重力定数で、6.67×10^{-11} Nm^2/kg^2）。地上の物体に働く重力（gravity）も、上の式から計算することができる。地上の物体の質量 m_1 を1 kg、地球の質量 m_2 を 5.97×10^{24} kg、地球の平均半径 r を 6.37×10^6 m として計算すると、$F = 9.81$ N となる。

地上すれすれの軌道（orbit）を周る人工衛星（satellite）が地面に落ちないようにするためには、水平方向に約 7.9 km/s の速度（第1宇宙速度：the first escape velocity）で飛び続ける必要がある。崖からボールを水平に投げるように、人工衛星も地球に対して常に落下し続けているが、地球は球体なので、第1宇宙速度に達したときに人工衛星は軌道を描いて地球を周回することになる。地球の重力は、高度が高くなるにつれて弱まる。赤道上空高度 36,000 m 付近を周っている人工衛星は、地球の自転速度と同じ 3.1 km/s で周っており、静止衛星（stationary satellite）と呼ばれる。地球の重力を脱して、太陽を周る軌道に乗るには秒速約 11.2 km/s（第2宇宙速度：the second escape velocity）で飛ぶ必要があり、さらに太陽系から脱して銀河系に出るためには、秒速約 16.7 km/s（第3宇宙速度：the third escape velocity）が必要になると考えられている。

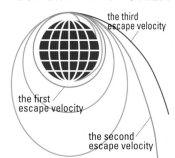

Exercise 1

[1] What Is Physics?

1) What is physics?
2) List some fields of study in classical physics.
3) List some fields of study in modern physics.
4) Who discovered the atomic nucleus?
5) What do you call the smallest particles that constitute all matter?

[2] Light

1) What is the relation between the angle of incidence and the angle of reflection?
2) What is refraction?
3) What is the speed of light?
4) Name the two types of lenses.
5) How do we perceive that something is red?

[3] Sound

1) What is the speed of sound through the air?
2) Name the two types of waves.
3) What are the three elements of sound?
4) What is the relation between wavelength and frequency?
5) What is amplitude and how does it affect sound?

[4] Force

1) What are the two types of forces?
2) What is the SI unit of force?
3) Name the three elements of force.
4) What is equilibrium?
5) What does the law of action and reaction (Newton's third law of motion) state?

[5] Motion

1) What does the law of inertia (Newton's first law of motion) state?
2) What is the equation of motion (Newton's second law of motion)?
3) What is gravity?
4) What is the gravitational acceleration value?
5) How far will a free falling object drop in 10 seconds?

A. Name the two types of images that are formed according to the position of the object in relation to the focal point.

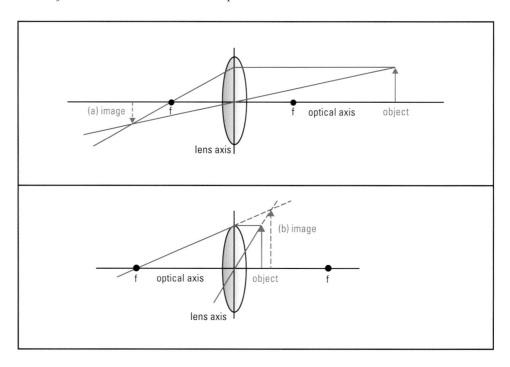

B. (1) Indicate wavelength and amplitude in the chart.
   (2) What is the frequency of this wave?

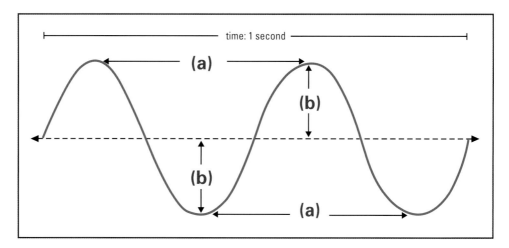

Electricity

静電気

static electricity

雷

lightning

自由電子

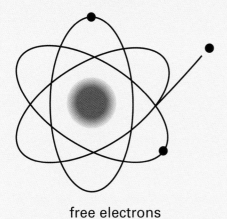

free electrons

電気回路

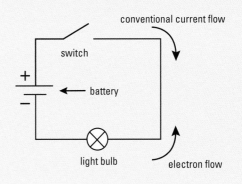

electric circuit

ⓘ Reading

CD-38

Static Electricity

If you rub a plastic straw with a piece of tissue paper, the straw and the tissue paper will **attract** each other. This phenomenon is caused by **static electricity**. When these things are rubbed together, friction makes **electrons** move away from the tissue paper to the straw, making the straw **negatively charged** (−) and the tissue paper **positively charged** (+). Opposite charges attract each other while like charges **repel** each other. Therefore, the straw and the tissue paper attract each other, while two straws rubbed by tissue paper repel each other (Fig. 1).

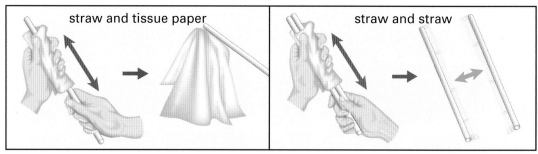

Fig. 1 Static Electricity

The **electric charge** caused by rubbing two different substances is called static electricity because it stays on each substance rather than flowing away. Whether a substance is negatively charged or positively charged depends on the material and the combination of the two substances. Between the two substances to be rubbed, those which are more likely to gain electrons will be negatively charged, and those which are more likely to lose electrons will be positively charged. A list of materials showing their tendency to gain or lose electrons is called the **triboelectric series** (Fig. 2).

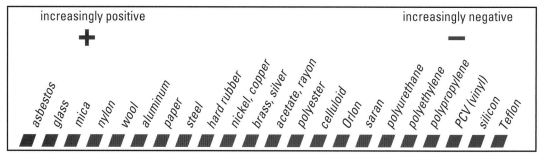

Fig. 2 Triboelectric Series

NOTES (p. 39)

1) powerは単位時間当たりの仕事量のことで、仕事率とも言う。したがって、以下の関係が成り立つ。仕事量（energy）＝仕事率（power）× 時間（time）

Current Electricity

Current electricity is the electricity that flows along a **conductor**. In other words, it is a constant movement of electrons. Although electrons actually move from negative to positive, we say **electric current** flows from positive to negative following tradition. We can see how an electric current flows by setting up an **electric circuit** which consists of a **wire**, **batteries**, a **switch**, and a **resistor** such as a light bulb. There are two kinds of electrical circuits: **series circuits** and **parallel circuits** (Fig. 3). In a series circuit, the electric current is equal through all components; while in a parallel circuit, the **voltage** is equal through all components.

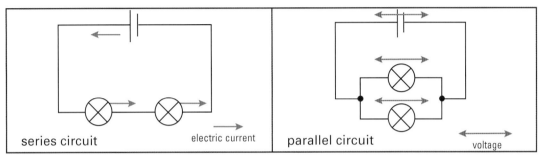

Fig. 3 Electric Circuit

Electric current is measured in **amperes** (A), which indicates the rate of electric flow in a circuit. One ampere (1 A) means a flow of 6.24×10^{18} electrons per second. Voltage is the force pushing electric current through wires. It is actually the potential difference of electric charge between two points, which is measured in **volts** (V). **Resistance** is a material's opposition to the flow of electric current, which is measured in **ohms** (Ω). **Ohm's law** states that the voltage equals the electric current times the resistance ($V = IR$, where V is voltage, I is electric current, R is resistance). For example, if the current is 20 A and the resistance is 5 Ω, the voltage must be 100 V (100 V = 20 A × 5 Ω).

Electric Power and Electric Energy

Electricity can do many things: light a bulb, generate sound, produce heat, turn motors, etc. Electricity's ability to do work is called **electric power**[1]. The SI unit of electric power is the **watt** (W), which is calculated by multiplying voltage and electric current ($P = VI$, where P is electric power, V is voltage, I is electric current). So, 1 W means the electric power from a current of 1 A flowing through 1 V (1 W = 1 V × 1 A). Since electric power is **electric energy** consumed per unit time, the total electric energy consumption is calculated by watt times hour (Wh). For example, if 100 W of electric power is used for 10 hours, the consumed electric energy is 1,000 Wh (100 W × 10 h), which is also expressed as 1 kWh.

Vocabulary

□ …を引きつける	attract	□ スイッチ	switch
□ 静電気	static electricity	□ 抵抗器	resistor
□ 電子	electron	□ 直列回路	series circuit
□ 負に帯電した	negatively charged	□ 並列回路	parallel circuit
□ 正に帯電した	positively charged	□ 電圧	voltage
□ 反発する	repel	□ アンペア	ampere (A)
□ 電荷	electric charge	□ ボルト	volt (V)
□ 帯電列	triboelectric series	□ 抵抗	resistance
□ 伝導体	conductor	□ オーム	ohm (Ω)
□ 電流	electric current	□ オームの法則	Ohm's law
□ 電気回路	electric circuit	□ 電力	electric power
□ 導線	wire	□ ワット	watt (W)
□ 電池	battery	□ 電力量	electric energy

Useful Expressions

1) Lightning (   ) (   ) (   ) static electricity in the clouds.
（雷は、雲の中の静電気<u>によって起こる</u>）

2) Whether a substance is negatively charged or positively charged (   )
(   ) the nature of the material.（ある物質が負に帯電するか、正に帯電
するかは、その物質の材質に<u>依存する</u>）

3) A plastic straw is more (   ) (   ) gain electrons than tissue
paper.（プラスチックのストローは、ティッシュペーパーよりも電子を<u>受け取り
やすい</u>）

⬢Experiment

□ Purpose: To measure voltage and electric
current through components in a series
circuit and a parallel circuit.

□ Hypothesis: (1) In a series circuit, electric
current is equal through all components,
but voltage is not. (2) In a parallel circuit,
voltage is equal through all components,
but electric current is not.

□ Apparatus and materials: two dry-cell
batteries (1.5 V each) connected in series (3
V), two miniature bulbs (10 Ω), conducting
wires, a switch, a voltmeter, an ammeter

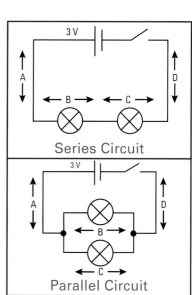

☐ Procedure:
- Set up a series circuit and a parallel circuit according to the circuit diagrams.
- Measure voltage and electric current between the designated points.

☐ Result: (example)

Series Circuit	A	B	C	D
Voltage (V)	3 V	(1)	1.5 V	3 V
Current (A)	0.15 A	0.15 A	(2)	0.15 A
Parallel Circuit	A	B	C	D
Voltage (V)	3 V	(3)	3 V	3 V
Current (A)	0.6 A	0.3 A	(4)	0.6 A

☐ Conclusion: Both hypotheses are supported.

Functional Expressions

1) (　　　) (　　　) a series circuit according to the circuit diagram.（回路図に従って、直列回路を組み立てなさい）

2) Measure voltage between the two (　　　) points.（指定された 2 点間の電圧を測りなさい）

◉Presentation

Slide　　　　　　　　　　　Prompter Card

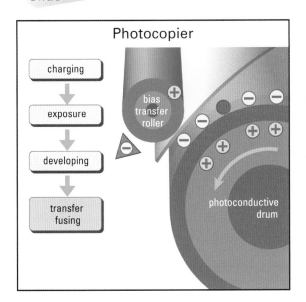

- A photocopier makes use of static electricity.
- Inside a photocopier, there is a "photoconductive drum."
- The photoconductive material conducts electricity when exposed to light.
- When a document is scanned, the paper's blank space reflects white light.
- The parts on the drum exposed to light lose the charge.
- The remaining parts attract toner and form images.

Q & A

Q. ドラムに付着したトナーを、どうやって紙に移すのですか。

A. 通常はトナーをプラスに帯電させ、マイナスに帯電させた紙に吸着させます。それを熱によって焼き付けます。

学習活動

1) 電気に関する現象、実験、製品などについてワンポイント・スライドを作成してみよう。

2) プロンプターカードを作成して、ミニ・プレゼンテーションをしてみよう。

3) 英語で質問や答えを作成して、互いに質疑応答の練習をしてみよう。

電気の狩人たち

Benjamin Franklin

Ampère

Ohm

アメリカの科学者であり、政治家でもあるベンジャミン・フランクリン（Benjamin Franklin 1706–1790）が、凧を使った実験で、雷の正体が静電気であることをつきとめたのは、1752 年のことである。1800 年には、イタリアのボルタ（Alessandro Volta 1745–1827）が電池を発明し、1820 年にフランスのアンペール（André-Marie Ampère 1775–1836）が電流の向きを＋から－と定めた。その後 1897 年に、イギリスのトムソン（Joseph John Thomson 1856–1940）が、陰極線（cathode ray）の研究によって、電流の正体は－の電荷を帯びた電子の流れであり、その向きは－から＋であることをつきとめたが、現在でも慣習にならって、電気は＋から－へ流れることになっている。オームの法則は、ドイツの物理学者オーム（Georg Simon Ohm 1789–1854）が 1827 年に発表した。電圧、電流、抵抗の単位は、それぞれボルタ、アンペール、オームを記念して定められたものである。また電力の単位であるワットは、蒸気機関の開発で有名なスコットランドの技術者ワット（James Watt 1736–1819）にちなんだものである。

Volta

Thomson

Watt

Electromagnetism

U字磁石

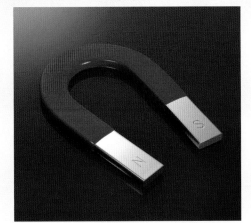

horseshoe magnet

棒磁石

bar magnet

電磁場

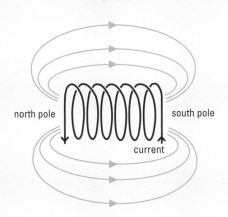

north pole · south pole · current

electromagnetic field

モーター

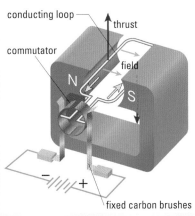

conducting loop · thrust · commutator · field · N · S · fixed carbon brushes · − · +

electric motor

◉ Reading

(Magnetic Fields)

A magnet attracts iron and attracts or repels other magnets. This force is called **magnetic force**. If you sprinkle **iron sand** over a bar magnet, a pattern of lines is produced (Fig. 1-a). These lines are called the **lines of magnetic force**, and they form a **magnetic field**. If you place a compass near a bar magnet, the pointer (the needle end marked "N") points to the S pole of the bar magnet (Fig. 1-b). In other words, a compass needle points in the direction of the magnetic force lines, which originate from the N pole and lead to the S pole. The earth acts like a giant magnet. A compass needle points to the north because the earth has its magnetic south pole in the north. It is rather confusing, but the magnetic poles of the earth do not match the geographic poles (Fig. 1-c). The earth's magnetic force is called **geomagnetism**.

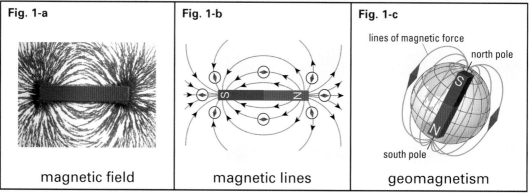

Fig. 1-a magnetic field   Fig. 1-b magnetic lines   Fig. 1-c geomagnetism

Fig. 1 Magnetic Field

(Electricity and Magnetic Fields)

When electric current flows in a wire, it generates a magnetic field around the wire. The direction of the magnetic field can be **determined** using your right hand. If you point your thumb in the direction of the current, your fingers will **curl** in the direction of the magnetic field. This is called the **right-handed screw rule** (Ampère's law) (Fig. 2-a). By winding the wire into a **coil** (solenoid), you can strengthen the magnetic field. Again, you can determine the direction of the magnetic field using your right hand. This time, curl your fingers in the direction of the current, then your thumb will point in the direction of the magnetic field (Fig. 2-b). This is another version of the right-handed screw rule.

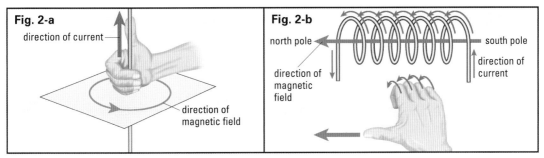

Fig. 2 Right-handed Screw Rule (Ampère's law)

Electromagnetic Induction

When electric current flows in a magnetic field, the magnetic field exerts a force on the electric current. This force is called **electromagnetic force**. An **electric motor** rotates because of this force. The direction of the electromagnetic force can be determined by **Fleming's left-hand rule**. If you hold your left hand with the **thumb**, **index finger**, and **middle finger** all at **right angles** to each other, the thumb represents the direction of the force, the index finger represents the direction of the magnetic field, and the middle finger represents the direction of the current (Fig. 3-a). It works in the opposite way with a **generator**. When a conductor (electric wire) moves in a magnetic field, electric current is generated. This is called **electromagnetic induction**, and the current thus generated is called **induced current**. The direction of induced current can be determined by **Fleming's right-hand rule**. If you hold your right hand with the thumb, index finger, and middle finger all at right angles to each other, the thumb represents the direction of the force, the index finger represents the direction of the magnetic field, and the middle finger represents the direction of the induced current (Fig. 3-b). When a bar magnet moves in and out of a coil, electric current flows in the coil. **Lenz's law** states that the induced current flows in such a way that the magnetic field produced by the current **obstructs** the motion of the magnet. So if a bar magnet (N pole) is **inserted** into a coil, a current is induced in such a way that opposes the incoming lines of magnetic force, and if the bar magnet is **withdrawn**, the current flows in the opposite way (Fig. 3-c).

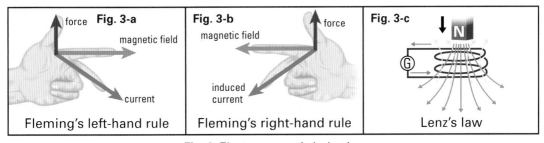

Fig. 3 Electromagnetic Induction

Vocabulary

☐ 磁力	magnetic force	☐ 親指	thumb
☐ 砂鉄（鉄粉）	iron sand (iron powder)	☐ 人差し指	index finger
☐ 磁力線	lines of magnetic force	☐ 中指	middle finger
☐ 磁界（磁場）	magnetic field	☐ 直角	right angle
☐ 地磁気	geomagnetism	☐ 発電機	generator
☐ 確定する	determine	☐ 電磁誘導	electromagnetic
☐ 巻く	curl		induction
☐ 右ねじの法則	right-handed screw	☐ 誘導電流	induced current
	rule (Ampère's law)	☐ フレミングの	Flemming's
☐ コイル	coil (solenoid)	右手の法則	right-hand rule
☐ 電磁（気）力	electromagnetic force	☐ レンツの法則	Lenz's law
☐ モーター	electric motor	☐ 妨げる	obstruct
☐ フレミングの	Flemming's	☐ 差し込む	insert
左手の法則	left-hand rule	☐ 引き抜く	withdraw

Useful Expressions

1) The lines of magnetic force (　　　) (　　　) the N pole and (　　　) (　　　) the S pole. （磁力線は、N 極から出て、S 極につながる）

2) Curl your fingers (　　　) (　　　) (　　　) (　　　) the current, then your thumb will point in the direction of the magnetic field. （指を電流が流れる方向に巻くと、親指が磁界の方向を指す）

3) Hold your left hand (　　　) the thumb, index finger, and middle finger all (　　　) right angles to each other. （左手の親指、人差し指、中指をすべて互いに直角になるようにしなさい）

◈Experiment

☐ Purpose: To observe how electric current flows when a bar magnet moves in and out of a coil.

☐ Apparatus and materials: a bar magnet, a coil, a galvanometer

☐ Procedure
- Connect a coil to a galvanometer.
- Move a bar magnet up and down in the coil and see how the needle of the galvanometer moves.

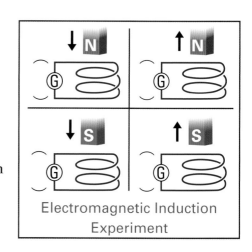

Electromagnetic Induction Experiment

☐ Result:

Magnetic pole	N	N	S	S
Movement of the magnet	↓	↑	↓	↑
Direction of the current	↓	↑	↑	↓

☐ Conclusion: The induced current flows in such a way that the magnetic field produced by the current opposes the motion of the magnet.

Functional Expressions

1) (   ) a coil (   ) a galvanometer. (コイルを<u>検流計につなぎなさい</u>)

2) (   ) a bar magnet (   ) and (   ) in the coil. (棒磁石をコイルの中で<u>上下に動かしなさい</u>)

⊙ Presentation

Slide

Prompter Card CD-52

Microwave Oven

- Microwaves are electromagnetic waves.
- Their wavelengths range from 1 m to 1 mm (longer than infrared rays / shorter than radio waves).
- Their frequency range is from 300 MHz to 300 GHz.
- A microwave oven produces microwave radiation at a frequency of 2.45 GHz.
- The microwaves oscillate water molecules and thus heat up water in food.
- Substances that do not contain water will not be affected.

Q. 電子レンジに金属や、アルミ・ホイルを入れるとどうなりますか。

A. 火花が散ったり、放電したりすることがあるので、危険です。

学習活動

1) 電磁気に関する現象、実験、製品などについて、ワンポイント・スライドを作成してみよう。

2) プロンプターカードを作成して、ミニ・プレゼンテーションをしてみよう。

3) 英語で質問や答えを作成して、互いに質疑応答の練習をしてみよう。

電磁気あれこれ

棒磁石（bar magnet）は、ふつう赤色（N 極）と青色（S 極）に塗り分けられているが、これを真ん中で 2 つに切っても、やはり片側が N 極で、もう一方が S 極になる。物質の中にある電子は、それ自体が N 極、S 極を持った小さな磁石で、通常はばらばらの方向を向いているが、これらの電子の磁極が同じ方向を向いた状態にあるのが磁石である。鉄くぎに磁石を近づけると、鉄釘の中の電子が一方向に並んで、磁石になる。鉄くぎのように磁石になりやすい性質を持った物質を、強磁性体（ferromagnetic body）という。コイルの中に鉄芯を入れて電流を流すと、コイルによる磁界によって、鉄芯が磁石となり、一層磁力が強まる。これが電磁石（electromagnet）である。

　電磁波（electromagnetic wave）は、電磁誘導（electromagnetic induction：電気と磁気の相互作用）に伴って派生する波動で、横波（transverse wave）として空間を伝わっていく。電磁波は真空中でも伝わり、基本的に直進するが、空気や水のような媒質（medium）を通る際には、反射・屈折・回折・吸収などが起こり、また強い重力場では曲がることが知られている。光も電磁波のひとつである。電磁波は波長の長いものから順に、電波、マイクロ波、赤外線、可視光線、紫外線、X 線、ガンマ線に分けられる。

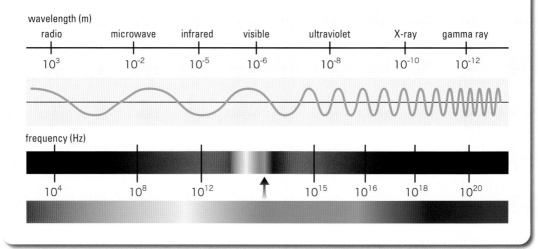

Unit
8

Work and Energy

ジェットコースター

roller coaster

仕事

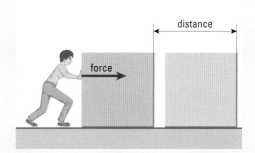

work

タービン

turbine

風力エネルギー

wind energy

CONTENTS

◉ Reading

Mechanical Energy

Energy is the ability to do **work**, or the ability to cause things to move, **deform**, break, etc. A moving object has energy because it has the ability to move, deform, or break other objects by collision. This energy is called **kinetic energy**[1]. An object is also considered to have energy when it is lifted up, because it has a potential to do work. This energy, on the other hand, is called **potential energy**. The higher the object is positioned, the larger its potential energy is. The potential energy is **converted** into kinetic energy as the object falls and accelerates under gravity.

Both kinetic energy and potential energy are called **mechanical energy**, and although they are **interchangeable**, their sum always remains the same. This principle is called the **law of conservation of mechanical energy**. For example, the potential energy of a pendulum increases as it moves up while its kinetic energy decreases. Conversely, the potential energy decreases as the pendulum moves down while its kinetic energy increases (Fig. 1).

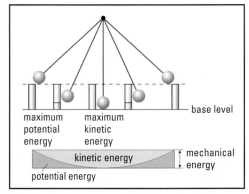

Fig. 1 Conservation of Mechanical Energy

Work and Energy

In physics, we say "work" is done when a force is applied to an object so that the object moves toward the direction of the force. In other words, the work is the **product** of force and distance (Work = Force × Distance). The SI unit of work is the **joule** (J), where 1 J of work means the work to move an object 1 m by 1 N of force (1 J = 1 Nm). So, if you move an object 2 m by a force of 50 N, the work is 100 J (50 N × 2 m = 100 J). Since energy is the ability to do work, the amount of energy can be measured by the amount of work. The potential energy (PE) of an object, therefore, equals the gravity acting on the object's mass times its position (height) from a **base level** ($PE = mgh$). For example, if an object of 2 kg is lifted up 3 m, the object has a potential energy of 60 J (approx.[2] 20 N × 3 m ≒ 60 J) (Fig. 2-a). On the other hand, kinetic energy (KE) is calculated by the following equation: $KE = \frac{1}{2}mv^2$ (m: kg / v: m/s). The equation means that kinetic energy is proportional to the mass of an object and the square of its speed. For example, the kinetic energy of a car (1 t = 1,000 kg) moving at 36 km/h (= 10 m/s) is 50,000 J [$\frac{1}{2}$ × 1,000 kg × (10 m/s)^2]. From the equation, we can **infer** that if the car moves twice as fast, the kinetic energy will be **quadrupled** and thus the braking distance will be four times longer (Fig. 2-b).

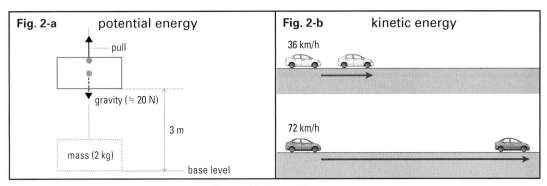

Fig. 2 Work and Energy

Energy Transformations

Energy has various forms—light energy, sound energy, electric energy, mechanical energy, thermal energy[3], chemical energy, etc.—and changes from one form to another. This is called **energy transformation** (Fig. 3). For example, kinetic energy can turn into electric energy, and electric energy can turn into thermal energy. Although energy may change its forms, it can neither be created nor destroyed; the total amount of energy always remains the same. This is called the **law of conservation of energy**. The SI unit of energy, as well as of work, is the joule (J). We have already seen how to measure mechanical energy, so let's look at how thermal energy is measured. Joule (1818–1889) found in his experiment that 4.2 J of kinetic energy produces 1 cal[4] of heat. Therefore, 1 J is **equivalent** to 0.24 cal of heat (1 J ≒ 0.24 cal). One joule (1 J) is also known to be equivalent to the heat produced when 1 W of electricity is used for 1 second (1 J = 1 Ws). So, if we use 500 W of electricity for 1 minute (= 60 s), the produced thermal energy is 30,000 J (500 W × 60 s).

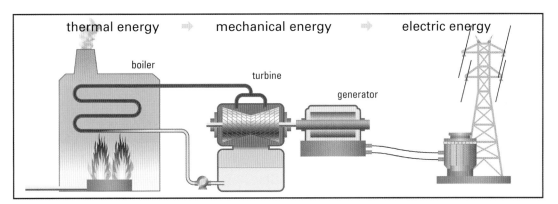

Fig. 3 Energy Transformations

NOTES

1) motion energyとも言う。 **2)** approx. (= approximately)「およそ・約」 **3)** heat energyとも言う。 **4)** 1 calは、1 gの水を1℃上昇させる熱量。

Vocabulary

□ 仕事	work	□ 積	product
□ 変形する	deform	□ ジュール	joule (J)
□ 運動エネルギー	kinetic energy	□ 基準面	base level
□ 位置エネルギー	potential energy	□ 推定する	infer
□ 変換する	convert	□ 4倍にする	quadruple
□ 力学的エネルギー	mechanical energy	□ エネルギー変換	energy transformation
□ 交換可能な	interchangeable	□ エネルギー保存の法則	law of conservation of energy
□ 力学的エネルギー保存の法則	law of conservation of mechanical energy	□ …に相当する	equivalent (to)

Useful Expressions

1) (   ) (   ) the object is positioned, (   ) (   ) its potential energy is. （物体の置かれる位置がより高いほど、その物体の位置エネルギーはより大きくなる）

2) The sum of kinetic energy and potential energy always (   ) (   ) (   ). （運動エネルギーと位置エネルギーの和は常に同じである）

3) Work is the (   ) (   ) force (   ) distance. （仕事は、力と移動距離の積である）

4) Kinetic energy is proportional to the mass of an object and the (   ) (   ) its speed. （運動エネルギーは、物体の質量と、速度の2乗に比例する）

5) Energy changes (   ) (   ) form (   ) (   ). （エネルギーは、次から次へと形を変える）

◆ Experiment

- □ **Purpose:** To observe the relation between kinetic energy and the mass / speed of an object.
- □ **Apparatus and materials:** steel balls (10 g, 20 g, 30 g), a wood cube, a speed sensor, a runway (with a column and a scale)
- □ **Procedure:**
  Experiment A: Let the three steel

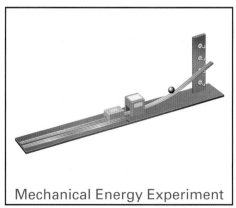

Mechanical Energy Experiment

52

balls roll down from the same height respectively and collide with the wood cube, and measure the distance the wood cube moves.

Experiment B: Let the same steel ball (for example, 20 g) roll down from different heights (for example, 10 cm / 20 cm / 30 cm) and collide with the wood cube, and measure the distance the wood cube moves.

☐ Result: (Example)

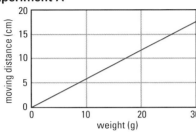

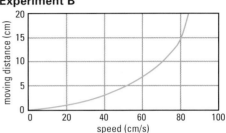

☐ Conclusion: Kinetic energy is proportional to the mass of an object and the square of its speed.

Functional Expressions

1) Let the three steel balls (　　　) (　　　) from the same height respectively.（3つの鋼球をそれぞれ同じ高さから転がしなさい）

2) (　　　) the (　　　) the wood cube moves.（木片が移動した距離を測りなさい）

◉ Presentation

Slide

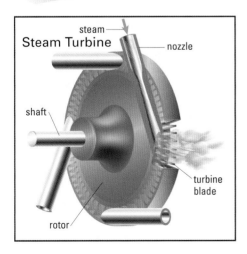

Prompter Card

- A steam turbine converts thermal energy into mechanical energy.
- It consists of a rotor and a shaft.
- High-pressure steam from the boiler spins the rotor at high speed.
- The rotary motion is transmitted by the shaft.
- The main use of steam turbines is electric power generation.
- The mechanical energy is further converted into electric energy.

Q & A

Q. タービンには、大きさの異なるローターが、幾重にも組み合わされているものがありますが、なぜですか。

A. 水蒸気は、圧力の低下とともに体積を増し、スピードが加速していきます。そこでローターは、下流に行くに従って羽根車を大きくし、水蒸気の運動エネルギーを効率よく取り込めるようにしてあります。

【学習活動】

1) エネルギーに関する現象、実験、製品などについて、ワンポイント・スライドを作成してみよう。

2) プロンプターカードを作成して、ミニ・プレゼンテーションをしてみよう。

3) 英語で質問や答えを作成して、互いに質疑応答の練習をしてみよう。

エネルギー七変化

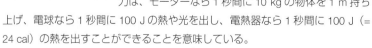

エネルギーは「仕事」をする能力なので、その大きさは「仕事量（＝力×移動距離）」で表すことができる。仕事量の単位はジュール（J）で、1 J は、1 N の力で、物体を 1 m 動かしたときの仕事量（＝ 1 Nm）を表す。これは、およそ質量 100 g の物体を 1 m 引き上げる仕事量に等しく、またその物体が、1 m 引き上げられた場所で持つ位置エネルギーに等しい。力学的エネルギー保存の法則（law of conservation of mechanical energy）によれば、この物体が 1 m 落下した時点で、先ほどの位置エネルギー（1 J）はすべて運動エネルギー（1 J）に置き換わっていることになる。ここで、運動エネルギーの公式（$KE = \frac{1}{2}mv^2$ (m: kg / v: m/s)）が成り立っているか、確かめてみよう。質量は 100 g だから、$m = \frac{1}{10}$ (kg)。この物体が 1 m 落下するのに要する時間は、重力加速度を約 10 m/s^2 とすると、$d = \frac{1}{2}at^2$ から、$1 = \frac{1}{2} \times 10 \times t^2 \therefore t = \sqrt{\frac{1}{5}} = \frac{\sqrt{5}}{5}$。したがって、その時点での速度は、$v = at = 10 \times \frac{\sqrt{5}}{5} = 2\sqrt{5}$ (m/s^2) となる。よって、$\frac{1}{2}mv^2 = \frac{1}{2} \times \frac{1}{10} \times (2\sqrt{5})^2 = 1$ (J) となり、確かに元の位置エネルギーと一致していることがわかる。実は運動エネルギーの公式（$KE = \frac{1}{2}mv^2$）は、位置エネルギーの公式（$PE = mgh$）から簡単に導くことができる。この物体が h m 落下したとき、$h = \frac{1}{2}gt^2$ だから、$mgh = \frac{1}{2}mg^2t^2 = \frac{1}{2}m(gt)^2$ となる。ここで $v = gt$ より、$\frac{1}{2}m(gt)^2 = \frac{1}{2}mv^2$。このように、位置エネルギーの公式（$PE = mgh$）と運動エネルギーの公式（$KE = \frac{1}{2}mv^2$）は等価である。

Joule

ジュール（James Prescott Joule 1818–1889）は、重りの力を使って水中の羽根車を回し、重りの運動と水温の上昇の関係を調べた結果、4.2 J の仕事が、水温を 1℃上昇させることをつきとめた（4.2 J = 1 cal）。また電力の単位であるワット（W）は、もともと「仕事率」の単位であり、1 秒間になしうる仕事の量（J/s）を表している。1 W は、たとえばモーターが質量 100 g（≒ 1 N）の物体を 1 秒間で 1 m 引き上げた場合の仕事に相当する（1 W = 1 N•m/s）。したがって、100 W の電力は、モーターなら 1 秒間に 10 kg の物体を 1 m 持ち上げ、電球なら 1 秒間に 100 J の熱や光を出し、電熱器なら 1 秒間に 100 J（= 24 cal）の熱を出すことができることを意味している。

Thermodynamics

熱気球

hot-air balloon

温度計

temperature

サーモグラフィー

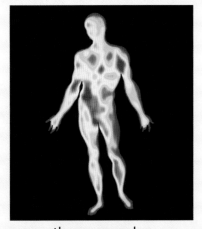

thermography

火力発電

thermal power generation

CONTENTS

Reading

Heat and Temperature

Heat changes the state of matter from **solid** to **liquid** to **gas** (Fig. 1). In solids, the **molecules** are bound tightly to each other so they can only vibrate but not move to another location. This makes solids keep their shape and volume.[1] In liquids, the molecules are connected loosely to each other and they move freely. This makes liquids keep their volume but not their shape. In gasses, the molecules are not connected to each other, and they can move around freely. Therefore, gases have neither definite shape nor definite volume. If you observe small particles suspended in a liquid or gas, you will see that they are moving **randomly**. This phenomenon, known as **Brownian movement**, is caused by the impact of free-moving molecules (Fig. 2). Heat is actually the kinetic energy of molecules, and as molecules get more heat, their movement becomes larger. **Temperature** is the measurement of heat energy. A **thermometer** is a device to measure temperature in various ways. For example, a **mercury** thermometer measures temperature using the **thermal expansion** of liquid mercury. The lower limit of temperature is zero **kelvin** (0 K = −273°C), also known as the **absolute zero point**, while there is no upper limit to temperature.

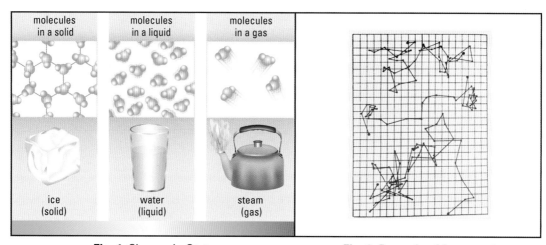

molecules in a solid

molecules in a liquid

molecules in a gas

ice (solid)

water (liquid)

steam (gas)

Fig. 1 Change in State

Fig. 2 Brownian Movement

Three Laws of Thermodynamics

The **first law of thermodynamics** is the law of the conservation of energy (see Unit 8), which states that the total amount of energy remains constant in an **isolated system**. Joule showed in his experiment in 1843 that mechanical work can turn into heat and thus heat energy can be measured by the amount of work done. However, he did not show the reverse would apply, or that the generated

Unit 9 Thermodynamics

heat could be transferred back into the same amount of work. Actually, it is not possible to convert heat completely into work in a cyclic process according to the **second law of thermodynamics**. This is because, once mechanical energy (**external energy**) is turned into heat (**internal energy**[2]), the random movement of numerous molecules gets out of control and cannot be restored to the original work again. It means that although the total amount of energy (external energy + internal energy) remains the same, thermal energy transfer is **irreversible** and part of the internal energy can never revert to the external energy. In other words, **available energy**, or so-called **exergy**, decreases in the process of energy transfer, while the amount of energy that is no longer available for mechanical work (**entropy**) increases. This leads to the conclusion that it is impossible to make a **perpetual motion machine**. The second law of thermodynamics can be restated as: "Entropy increases over time." The increase of entropy means that things naturally change from order to disorder, in other words, things decay over time (Fig. 3). The **third law of thermodynamics** states that entropy will approach a minimum value as a system approaches the absolute zero point. From the second law and the third law, we can understand that heat will cause things to decay, and if there is no heat, things will never change. However, things can actually never reach the absolute zero point, because they can never be in contact with anything that is below the absolute zero point (Fig. 4).

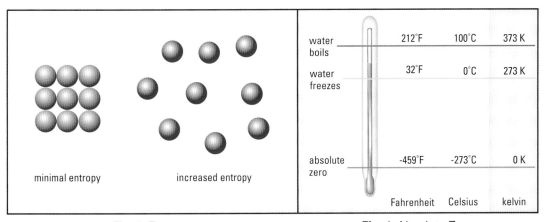

Fig. 3 Entropy　　　　　　　**Fig. 4** Absolute Zero

NOTES

1) 通常固体の分子同士のすき間は液体よりも詰まっているが、水の場合は固体の方がすき間が大きい。そのため、同じ質量では、固体（氷）の方が液体（水）よりも体積が大きくなる。
2) 分子の熱運動によるエネルギーと、分子相互間の位置エネルギーなどを加えたものを（物質の）内部エネルギーと呼ぶ。

Vocabulary

☐ 固体	solid	☐ 孤立系	isolated system
☐ 液体	liquid	☐ 熱力学の	second law of
☐ 気体	gas	第2法則	thermodynamics
☐ 分子	molecule	☐ 外部エネルギー	external energy
☐ 不規則に	randomly	☐ 内部エネルギー	internal energy
☐ ブラウン運動	Brownian movement	☐ 不可逆的な	irreversible
		☐ 利用可能なエネルギー	available energy
☐ 温度	temperature		
☐ 温度計	thermometer	☐ エクセルギー	exergy
☐ 水銀	mercury	☐ エントロピー	entropy
☐ 熱膨張	thermal expansion	☐ 永久機関	perpetual motion machine
☐ ケルビン	kelvin (K)		
☐ 絶対零度	absolute zero point	☐ 熱力学の	third law of
☐ 熱力学の 第1法則	first law of thermodynamics	第3法則	thermodynamics

Useful Expressions

1) Gases have (　　　) definite shape (　　　) definite volume.（気体には、特定の形<u>も</u>、特定の体積<u>もない</u>）

2) This (　　　) (　　　) (　　　) (　　　) that it is impossible to make a perpetual motion machine.（このことは、永久機関を作ることはできないという<u>結論を導く</u>）

3) The second law of thermodynamics (　　　) (　　　) (　　　) (　　　): "Entropy increases over time."（熱力学の第2法則は、「エントロピーは増大する」と<u>言い換えることができる</u>）

⊕ Experiment

☐ Purpose: To find the relation between the temperature, pressure, and volume of air.

☐ Apparatus and materials: a ring stand, a clamp, a syringe, a cap, weights, water, a container, a gas burner, a thermometer

☐ Procedure:

Experiment A: Assemble the apparatus as demonstrated. Set the initial position of the piston (= the volume of air). Place a weight

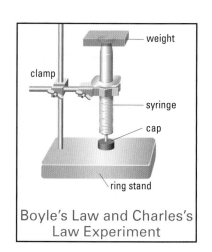

Boyle's Law and Charles's Law Experiment

on the piston and record the volume. Repeat the process adding an extra weight each time.

Experiment B: Put the syringe in water warmed to room temperature. Increase the water temperature and measure the volume of air.

☐ Result: (example)

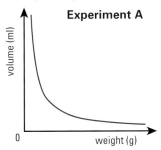

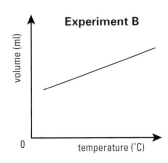

☐ Conclusion: At the same temperature, the volume of air is reversely proportional to the pressure, while under the same pressure, the volume of air is proportional to the temperature.

Functional Expressions

1) (　　　　) the apparatus as (　　　　). （実験装置を表示の通り組み立てなさい）

2) (　　　　) the (　　　　) (　　　　) of the piston. （ピストンの測定開始位置を設定しなさい）

⊙Presentation

Slide

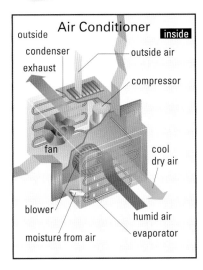

Prompter Card

- An air conditioner works in much the same way as a refrigerator.
- It uses a fluorocarbon refrigerant.
- It has three main components: an evaporator, a compressor, and a condenser.
- The refrigerant absorbs heat from air as it evaporates, and the cooled air is circulated back to the room.
- The refrigerant gas is compressed and becomes a hot high pressure gas.
- The heat is blown away by a fan and the refrigerant gas is cooled and condensed to a liquid and sent back to the evaporator.

Q. エアコンは、部屋の温度を下げるので、エントロピーが減少するように見えますが、これは熱力学の第2法則と矛盾しないのですか。

A. 部屋の中だけを見れば、温度は下がってエントロピーが減少したように見えますが、部屋の外には熱が放出されているので、全体として見ればエントロピーは減少しません。

学習活動

1) 熱に関する現象、実験、製品などについて、ワンポイント・スライドを作成してみよう。

2) プロンプターカードを作成して、ミニ・プレゼンテーションをしてみよう。

3) 英語で質問や答えを作成して、互いに質疑応答の練習をしてみよう。

ボイル・シャルルの法則

Boyle

Charles

ボイル（Robert Boyle 1627–1691）は「温度が一定のとき、気体の圧力（P）と体積（V）は反比例する」ことを発見した（ボイルの法則 Boyle's law）。これは、体積が小さくなれば、それだけ気体の密度が高くなって、単位面積あたりに分子がぶつかる回数が増え、逆に体積が大きくなれば、気体の密度が低くなって、分子の衝突回数が少なくなるためである。

一方、シャルル（Jacques Charles 1746–1823）は「圧力（P）が一定のとき、気体の体積（V）は、温度（T）に比例する」ことを発見した（シャルルの法則 Charles's law）。これは、温度が高くなると、気体の分子運動が激しくなることを示している。ボイルの法則（$PV =$ 一定値）と、シャルルの法則（$\frac{V}{T} =$ 一定値）の2つの式を合わせると、$\frac{PV}{T} =$ 一定値となり、そこから $PV = RT$ という式が得られる。これを気体の状態方程式（gas state equation）と呼び、R を気体定数（gas constant）と呼ぶ。気体は温度が下がると体積が減少していく。実在の気体では、

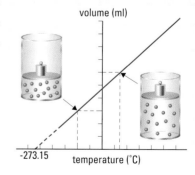

ある温度になると、凝縮（condensing）・凝固（freezing）が起こって液体や固体になるが、このような状態変化を起こさないような架空の気体（理想気体 ideal gas）を想定すれば、−273.15℃で気体の体積はゼロとなる。これが絶対零度（absolute zero point）である。

From Atom to Cosmos

核力

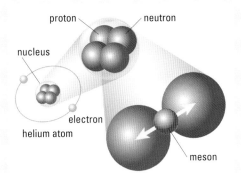

nuclear force

放射線

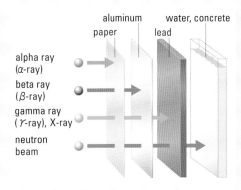

radiation

タイムトラベル

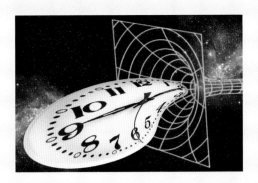

time travel

銀河

galaxy

CONTENTS

⊙Reading

Nuclear Physics

Although the **atom** is no longer the smallest particle as it used to be thought, it is still the basic unit of all things since it is the smallest component that retains the properties of an **element**. An atom is composed of electrons and a **nucleus**, and the nucleus consists of **protons** and **neutrons** (Fig. 1). Electrons are negatively charged and they orbit around the nucleus, which is positively charged. The atom has actually a very hollow structure if you consider the distance from the electron to the nucleus. If we compare the nucleus to the sun in the solar system, the **outermost electron** will be orbiting much farther away than **Neptune**.[1] The nucleus is held together by the strong interaction (**nuclear force**), which means the nucleus has a large potential energy. Einstein's formula $E = mc^2$ indicates that even a small mass may be converted into a tremendous amount of energy.[2] When there is any change in the nucleus, part of the energy is released as **radioactive rays**, which are either fast-moving particles or an electromagnetic wave. For example, the **alpha ray** (α-ray) is the helium nucleus (2 protons + 2 neutrons), the **beta ray** (β-ray) is an electron, and the **gamma ray** (γ-ray) is an electromagnetic wave. **Radioactive elements** have a rather unstable nucleus that decays over time and become more stable atoms or **isotopes** while emitting radioactive rays.[3] Uranium is a radioactive element which has a very unstable nucleus. When uranium-235 (U-235) is bombarded with neutrons, the nucleus is divided into nuclei of barium (Ba) and krypton (Kr) and emits several neutrons. This nuclear reaction is called **nuclear fission** (Fig. 2). The emitted neutrons will then bombard with other nuclei, which causes a **chain reaction** producing enormous amounts of energy. There is another nuclear reaction called **nuclear fusion**, in which multiple atomic nuclei are combined to form a heavier nucleus. This happens in the sun where hydrogen nuclei combine into helium nuclei, giving off large amounts of energy.[4]

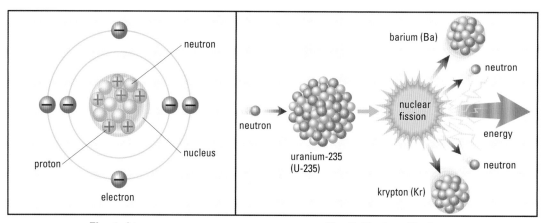

Fig. 1 Atom   **Fig. 2** Nuclear Fission

Theory of Relativity

 CD-70

Einstein's **theory of relativity** was a Copernican revolution to the Newtonian mechanics that was based on the ideas of absolute space and time. Although Newtonian mechanics was valid for almost every earthly phenomenon, scientists knew that it did not apply to events on the atomic scale or on the cosmic scale, and therefore, a new theory was needed to fill the gap. Einstein integrated space and time based on the principle that the speed of light is the same everywhere in the universe and nothing can travel faster than light. There are two theories of relativity: the **special theory of relativity** and the **general theory of relativity**. The special theory of relativity applies to inertia systems where no acceleration is involved. The theory holds that if person A travels away from B at a **sub-light speed**, A's time appears to flow slowly from B's point of view. From A's point of view, however, it works the other way around. For person A, B appears to move away from A at a sub-light speed so B's time must appear to flow slowly. This is called the **twin paradox** (Fig. 3), but the **time dilation** does occur when acceleration or gravity is involved.[5] The general theory of relativity holds that when an object accelerates, its mass becomes larger while space is **contracted** and time slows down (Fig. 4). At the speed of light, the mass is supposed to become infinitely large and time almost stops. This means nothing can actually reach the speed of light, which has no mass.

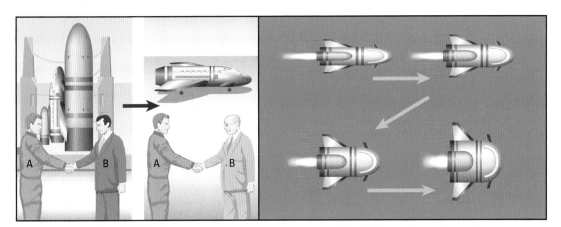

Fig. 3 Twin Paradox   **Fig. 4** Space-time Distortion

NOTES

1) 原子核の直径は約10^{-15} mで、原子の直径（約10^{-10} m）の約10万分の1である。
2) エネルギー（E）は、質量（m）と光速の2乗（c^2）に比例する。
3) ウラン（U）の原子番号は92で、陽子が92個含まれている。天然ウランのほとんどは、中性子が146個のU-238であるが、中性子が143個の同位体U-235も0.7％ほど存在する。
4) 正確には水素の同位体である重水素H-2の原子核（陽子1中性子1）が2つ結合してヘリウム原子（陽子2中性子2）となる。
5) カーナビゲーションで用いられているGPS（Global Positioning System）では、実際に相対性理論に基づいて、地上とGPS衛星の間に生じる時間のずれを計算して調整している。

Quantum Mechanics

Quantum mechanics is the study of matter and energy on atomic and subatomic levels. Subatomic particles include **nucleons** and **elementary particles** (Fig. 5). Nucleons are protons and neutrons, which are made of elementary particles called **quarks**. There is another group of elementary particles called **leptons**, which include electrons. Quarks and leptons are categorized as **fermions**, while there is yet another category of elementary particles called **bosons**, which work as agents to transmit the fundamental forces: The electromagnetic force is transmitted by **photons**, the weak interaction by W and Z bosons, and the strong interaction by **gluons**. At the level of elementary particles, the distinction between particles and waves becomes ambiguous, and their existence can only be quantified as a unit called a **quantum**. Quantum is the Latin word for "amount," and it means the smallest possible **discrete** unit of any physical property, such as matter or energy. Some elementary particles have masses while others do not, and when they interact with each other, energy is produced while the mass may be lost. The minimum unit of energy is called energy quantum, and the general amount of energy is always expressed as an **integral multiple** of a quantum.[6] The interactions of elementary particles are the key to understanding the generation of matter and energy. With the theory of relativity and quantum mechanics, modern physics is approaching the fundamental questions concerning the generation of matter and energy as well as the creation of the whole universe.

Fermion		Boson
<quark>	<lepton>	γ : photon
u: up	e: electron	W: W boson
c: charm	μ: muon	Z: Z boson
t: top	τ : tau	g: gluon
d: down	νe: electron neutrino	
s: strange	$\nu\mu$: muon neutrino	Higgs boson
b: bottom	$\nu\tau$: tau neutrino	(unidentified)

Fig. 5 Elementary Particles

NOTES

6) 例えば、光のエネルギー単位は$E = h\nu$（E: 単位エネルギー　h: プランク定数　ν ［ニュー］: 光の振動数）で表され、一般に光エネルギーはその整数倍（$En = nh\nu$）で表される。このように、ある物理量が連続的な値をとらず、ある単位量の整数倍で表される不連続な値しかとらない場合、この単位量を量子と呼ぶ。

Vocabulary

☐ 原子	atom	☐ 一般相対性理論	general theory of relativity
☐ 元素	element		
☐ 原子核	nucleus (*pl.* nuclei)	☐ 亜光速	sub-light speed
☐ 陽子	proton	☐ 双子のパラドクス	twin paradox
☐ 中性子	neutron		
☐ 最外殻電子	outermost electron	☐ 時間のずれ	time dilation
☐ 海王星	Neptune	☐ 収縮する	contract
☐ 核力	nuclear force	☐ 量子力学	quantum mechanics
☐ 放射線	radioactive ray	☐ 核子	nucleon
☐ アルファ線	alpha ray	☐ 素粒子	elementary particle
☐ ベータ線	beta ray	☐ クォーク	quark
☐ ガンマ線	gamma ray	☐ レプトン	lepton
☐ 放射性元素	radioactive element	☐ フェルミ粒子	fermion
☐ 同位体	isotope	☐ ボゾン粒子	boson
☐ 核分裂	nuclear fission	☐ 光子	photon
☐ 連鎖反応	chain reaction	☐ グルオン	gluon
☐ 核融合	nuclear fusion	☐ 量子	quantum
☐ 相対性理論	theory of relativity	☐ 不連続の・離散的な	discrete
☐ 特殊相対性理論	special theory of relativity	☐ 整数倍	integral multiple

Useful Expressions

1) The nucleus (   ) (   ) protons and neutrons.（原子核は、陽子と中性子<u>から成る</u>）

2) Electrons are (   ) (   ) while the nucleus is (   ) (   ).（電子は<u>マイナスに帯電しており</u>、一方原子核は、<u>プラスに帯電している</u>）

3) Newtonian mechanics is (   ) (   ) the ideas of absolute space and time.（ニュートン力学は、絶対空間および絶対時間という考え方<u>に基づいている</u>）

4) Einstein (   ) space (   ) time.（アインシュタインは、時間<u>と</u>空間<u>を統一した</u>）

5) Protons and neutrons are (   ) (   ) quarks.（陽子と中性子は<u>クォークでできている</u>）

⊕ Experiment

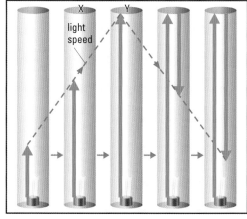

Light Clock

- ☐ Purpose: To understand why time appears to go slower in an object traveling at the speed of light according to the special theory of relativity. (Thought Experiment)
- ☐ Premise: The speed of light is the same for all observers regardless of their relative motion.
- ☐ Postulate:
  - Imagine a light clock that ticks and tocks at every second. The clock has a light source and two mirrors, one at the top and the other at the bottom. The clock "ticks" when the light pulse reaches the mirror at the top and "tocks" when it reaches the mirror at the bottom.
  - If the light clock were moving at the speed of light, how would it tick and tock from the external observer's point of view?
- ☐ Inference:
  - If the clock moves at the speed of light, the clock is meant to be at position Y in one second, when it is supposed to tick.
  - The light of the clock, however, appears to travel diagonally upward from the external observer's point of view.
  - Since the light travels on the diagonal path, which is longer than the horizontal path, it appears to be in the clock at position X. So, from the external observer's point of view, the light appears not to have reached the mirror at the top and therefore it does not appear to have ticked yet.
- ☐ Logical Consequence:
  - Time in an object traveling at the speed of light appears to go slower from the external observer's point of view.

Functional Expressions

 CD-74

1) (　　　　) a light clock that ticks and tocks at every second. （1 秒ごとに時を刻む光時計を想像しなさい）

2) The clock is (　　　　) (　　　　) be at position Y in one second. （時計は、1 秒後には、Y の位置にあるはずである）

3) The light of the clock appears to travel (　　　　) (　　　　) from the external observer's point of view. （時計の光は、外部の観察者の視点からは、斜め上方に進むように見える）

⊙ Presentation

Slide

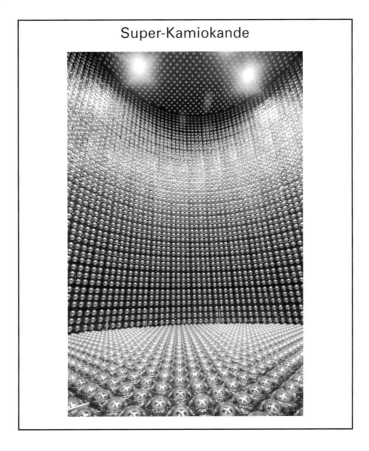

Super-Kamiokande

Prompter Card

- Super-Kamiokande (the Super-K) is a facility to observe elementary particles.
- It is a huge tank filled with ultrapure water, with thousands of photomultiplier tubes (PMTs) attached to its walls.
- It is located 1,000 m underground at Kamioka mine in Gifu prefecture.
- "Kamiokande" is short for Kamioka Neutrino Detection Experiment.
- When a neutrino interacts with electrons in water, it produces Cherenkov radiation, which can be recorded by the PMTs.
- Kamiokande, the prior facility, detected neutrinos produced by a supernova explosion.
- For this discovery, Masatoshi Koshiba was awarded the Nobel Prize in Physics in 2002.

Q. カミオカンデや、スーパーカミオカンデには、どうして大量の水が必要なのですか。

A. ニュートリノは非常に小さいために、ほとんどの物質を通り抜けてしまいますが、まれに他の物質と衝突することがあります。カミオカンデでは 3,000 トン、スーパーカミオカンデでは 50,000 トンの超純水を用意して、宇宙線に含まれるニュートリノと水の分子の衝突を観測しているのです。

学習活動

1) エネルギーに関する現象、実験、製品等について、ワンポイント・スライドを作成してみよう。

2) プロンプターカードを作成して、ミニ・プレゼンテーションをしてみよう。

3) 英語で質問や答えを作成して、互いに質疑応答の練習をしてみよう。

量子論の生みの親たち

光の正体をめぐる探究が量子力学を生むきっかけとなったことは、Unit 2 のコラムで紹介した。その先鞭をつけたのが「量子論の父」と呼ばれるプランク（Max Planck 1858–1947）である。プランクは溶鉱炉の熱と光の色の関係を調べていて、光のエネルギー（En）がある単

Planck

Einstein

位量（hV）の整数倍（nhV）となる離散的な値（discrete quantity）をとることを発見し、その単位量を量子（quantum）と名付けた。この発見に基づいてアインシュタイン（Albert Einstein 1879–1955）は、光がエネルギーを持った小さな粒子の集まりであると考え、それを光量子（light quantum）と名付けた（光子：photon とも呼ばれる）。アインシュタインは、光量子仮説を用いて光電効果（photoelectric effect：物質が光を吸収して自由電子を発生する現象）を説明し、この研究によって、ノーベル賞を受賞する。その後ラザフォード（Ernest Rutherford 1871–1937）によって原子核が発見されると、ボーア（Niels Henrik David Bohr 1885–1962）は、電子の軌道を量子化した原子モデルを発表し、電子は持っているエネルギーに応じて決められたとびとびの軌道（電子核：electron-shell）を周っていて、内側の軌道から外側の軌道に飛び移るときには光を吸収し、逆に外側の軌道から内側の軌道に飛び移るときには光を発するとした（$Eb-Ea = hV$ Eb：外側の軌道の電子エネルギー　Ea：内側の軌道の電子エネルギー）。ド・ブロイ（Louis de Broglie 1892–1987）は、電子軌道がとびとびになる理由を電子が波のようにふるまうからであると考えた。つまり電子が軌道を飛び続けるためには、電子の波が打ち消し合わないように、軌道の長さが電子の波長の整数倍になっていると考えたのである。さらにド・ブロイは、電子にかかわらずすべての物質を波ととらえる考え方（物質波：matter wave）を発表し、これが量子論の基本的なものの見方となっ

Rutherford

de Broglie

Bohr

Schrödinger

た。シュレーディンガー（Erwin Schrödinger 1887–1961）は、物質波の伝わり方を表す波動方程式（wave equation）を完成させ、ハイゼンベルグ（Werner Heisenberg 1901–1976）は、不確定性原理（uncertainty principle）を発表して、素粒子の位置と運動量を同時に確定することはできないとした。一方、素粒子レベルの物質のふるまいが確率論的にしか決まらないという考え方には反論もあり、量子力学のさらなる探究が続いている。

Heisenberg

Exercise 2

[6] Electricity

1) Name the two types of electricity.
2) Name the two types of electric circuits.
3) What does Ohm's law state?
4) What is electric power?
5) What is electric energy?

[7] Electromagnetism

1) Indicate the direction of magnetic force lines.
2) What does the right-handed screw rule state?
3) What does Fleming's left-hand rule state?
4) What is electromagnetic induction?
5) What does Lenz's law state?

[8] Work and Energy

1) Name the two types of mechanical energy.
2) What does the law of conservation of mechanical energy state?
3) What is the SI unit of work?
4) Calculate the kinetic energy of an object weighing 10 kg moving at a speed of 10 m/s.
5) What does the law of conservation of energy state?

[9] Thermodynamics

1) What is the difference between heat and temperature?
2) What does the first law of thermodynamics state?
3) What does the second law of thermodynamics state?
4) What is entropy?
5) What does the third law of thermodynamics state?

[10] From Atom to Cosmos

1) What is an atom composed of?
2) List some types of radioactive rays.
3) Name the two types of nuclear reaction.
4) What are the two theories of relativity called?
5) What is a quantum?

A. Calculate the amperage, voltage, or resistance at each designated point in the following circuit.

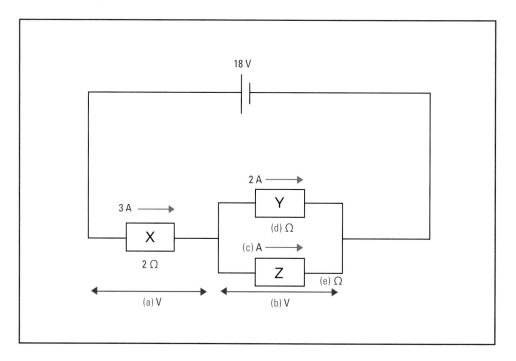

B. Indicate the direction of induced current in each case.

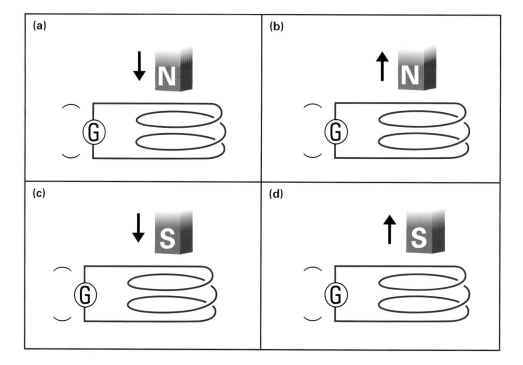

Appendix 1

Presentation Aids

1 学習活動

テキストの Presentation エクササイズを参考にして、グループで研究発表や製品紹介などの、プレゼンテーションをしてみよう。

(準備)

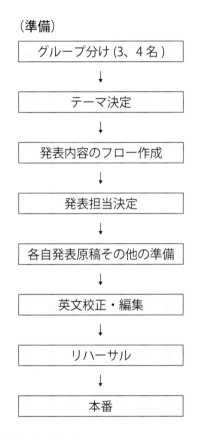

グループ分け (3、4名)
↓
テーマ決定
↓
発表内容のフロー作成
↓
発表担当決定
↓
各自発表原稿その他の準備
↓
英文校正・編集
↓
リハーサル
↓
本番

2 用意するもの

A. ハンドアウト

ハンドアウトは、視聴者の理解を助けるために配るものです。実際には発表スライドを縮小したものや、発表内容を要約したものが使われますが、学習活動では発表のアウトラインに絞ったハンドアウトを、以下のような点に留意して作成してみましょう。

- 発表の構成や流れがひと目で分かるような章立てにする
- 各章のポイントとなる項目を個条書きにする
- 参考資料（データなど）が必要な場合は、別途添付する
- ☞【ハンドアウトの例】(p. 77)

B. スピーチ原稿

スピーチ原稿は、あくまでも練習用です。暗唱できるまで何度も練習して、本番ではプロンプター・カードを使うようにしましょう。練習が効果的にできるように、以下のような工夫をしてみましょう。

- 左3分の1は、アウトライン用に空けておき、発表文は右側に書く
- ダブルスペースにして、行間に気がついたことを書き込めるようにする
- ポーズ（//）や、強調（太字・下線）など、各自で練習用のマークを工夫する
- ☞【スピーチ原稿の例】(p. 78)

C. プロンプター・カード

プロンプター・カードは、記憶の助けとなるように、必要なポイントのみ個条書きにしたものです。プロンプター・カードの最大の利点は、視聴者とのアイ・コンタクトを妨げないことです。

- プロンプター・カードは、市販の情報カード（B6版）を使う
- 必要なポイントのみ個条書きにする
- スライドとの同期など、必要に応じてマークをつける
- 視聴者の視線がスライドに集中するような場面では、原稿文を貼りつけておいて、それを読み上げてもよい（ただし、練習のために、できるだけ原稿に頼らないようにする）
- ☞【プロンプター・カードの例】(p. 79)

D. ビジュアル・エイド

発表を効果的にするために用いる小道具（props）には、以下のようなものがあります。

1) 実物（実演用）
2) ポスター、チャート、グラフなど
3) 動画、音など
4) 教材提示装置
5) プレゼンテーション・スライド

この中で最も一般的に用いられているのが、プレゼンテーション・スライドです。もちろんプレゼンテーション・スライドには、チャートやグラフのほか、動画や音なども埋め込むことができます。以下のような点に留意して、プレゼンテーション・スライドを作成してみましょう。

- 必要なポイントのみ個条書きにする（× 文字の多すぎるスライド）
- 図表はわかりやすく（× 数字の羅列 → ○ グラフによる視覚化）
- 虚飾に走らない（× 過度のアニメーション）
- ☞【プレゼンテーション・スライドの例】(p. 79)

❸ 評価

発表のときには、次のような評価項目についてピア・レヴューをしてみましょう。

Topic　発表内容について

1) Contents 内容の充実度
- 十分調べられ、興味の持てる内容になっていたか

2) Organization 構成
- わかりやすく、論理的な構成になっていたか

Delivery　発表技術について

3) English 英語
- 正しい英語を話せていたか（発音・文法）

4) Poise 落ち着き
- 声は聞き取りやすかったか
- アイ・コンタクトはとれていたか（☞注1）
- 立ち位置や、身振り手振りは自然だったか（☞注2）

5) Persuasion 説得性
- わかりやすかったか
- ビジュアル・エイドは効果的だったか

（注 1）Cast–Hook–Release Rule

効果的なアイ・コンタクトの取り方に、Cast–Hook–Release Rule というものがあります。つまり、まず釣り竿を投げる（cast）ように視線を投げかけ、次に魚がかかったら釣り上げる（hook）ように聴衆の 1 人の視線をとらえ、しばらくしたら魚を解き放つ（release）ように視線も離します。これを違った方向へ、まんべんなく繰り返していけばよいわけです。

（注 2）Speaker's Triangle

アイ・コンタクトと連動して、自然な立ち位置を確保する方法に Speaker's Triangle というものがあります。目の前に、図のような逆三角形をイメージして、辺に沿うように適宜移動しながら話す方法です。

（**audience**）

speaker

☞ 【評価シートの例】（p. 80）

❹ 質疑応答について

質疑応答は、聞き手にとってはプレゼンテーションの内容をより深く理解するための、また話し手にとってはプレゼンテーションの内容をよりよく理解してもらうための重要な機会です。しかし、英語母語話者でないわれわれにとって、最も難しいものでもあります。それだけにどのように質問し、どのように応答するのかをよく研究し、練習する必要があります。

▶ 質問の仕方

1) 質問の種類

質問には、大きく分けて Yes か No で答えられる Yes-No question (closed question) と、疑問詞で始まる WH question (open question) の2種類があります。質問の目的に応じて使い分けましょう。

2) 質問の目的

① 明確化（clarification）

聞き逃した個所や、はっきりしない部分について、言い直しや、言い換えを求める質問

（例）

- Sorry <u>I missed</u> the term you used to describe X. <u>Could you repeat it again</u>?
- <u>I'm not quite sure</u> what you meant by Y. <u>Can I ask you to paraphrase it</u>?

② 確認（confirmation）

自分の理解が正しいかどうか、確認するための質問

（例）

- <u>Do I understand that</u> your hypothesis is based on X theory?
- So, what you mean to say is X equals Y. <u>Am I correct</u>?

③ 情報要求（asking for information）

プレゼンテーションでは触れられなかった事例（example）やデータ（data）について、追加情報を求める質問

（例）

- 疑問詞で始まる疑問文全般（who / what / when / where / how much / how many）
- <u>Could you give me an example</u>?

④ 説明要求（asking for an explanation）

さらに詳細な説明や、特に因果関係について論理的な説明を求める質問

（例）

- So, you are claiming that X is more effective than Y. <u>Can you elaborate</u>?
- You concluded X is a result of Y. <u>Why</u>?

⑤ 反対尋問（cross examination）
反論がある場合に、相手を論駁するための質問
（例）

- I don't quite agree with you on X, because the result of another experiment supports Y. Do you have anything to say about the data that support Y?
- I think you are jumping to conclusions because the correlation doesn't support any inherent causality. Can you explain why X is inherently caused by Y?

3）質問のステップ

非母語話者による質問は、しばしば単発的な Yes-No question（closed question）や、茫洋とした WH question (open question) に終始する傾向があるようです。良い質問は、理解を深めるのに役立つ良い回答を引き出します。良い回答を引き出す質問には、次のようなステップがあります。

Step 1　謝辞（appreciation）
まずは発表者の労をねぎらい、質問に答えやすいムードを作る
（例）

- Thank you very much for your very informative presentation on X. That was quite impressive.

Step 2　質問の方向づけ（orientation）
自分がどのようなことについて関心を持っており、どのような点について聞きたいのか、質問の方向づけを与え、回答者に考える余裕を与える
（例）

- I am particularly interested in X because I myself was involved in . . . So my question is about . . .

Step 3　質問（question）
Step 2 の文脈に沿って、ピンポイントの質問をする（前述の各質問例を参照）

▶ 応答の仕方

1）応答の準備

質疑応答は、プレゼンテーションの内容をよりよく理解してもらうための絶好の機会です。また限られた時間内で十分説明できなかったことを補足する機会でもあります。まずは、これだけ調べたのだから、自分のほうが聞き手よりもよく知っているはずだという自信を持つこと。次に、せっかく苦労して調べたことをもっとよく知ってもらいたいという気持ちを持つことです。その上で、30 ～ 50 問程度の想定問答を作って練習をしておけば、準備は万全です。

2) 応答のステップ

  | Step 1 | 謝辞（appreciation）

  まずは、質問者の関心に対して感謝する

  （例）

  - Thank you for your question.
  - Thank you for your interest in X.

  | Step 2 | 明確化（clarification）

  質問の内容や意図がはっきりしない場合には、問い返す

  （例）

  - So you are asking about the possible side effects of X. Is that correct?
  - I don't exactly get your question. Do you want to know the correlation between X and Y?

  | Step 3 | 回答（answer）

  できるだけ的確な回答を心がけ、質問者の期待に応え得たかどうか確認をする

  （例）

  - Does that answer your question?
  - Did I answer your question?

3) 答えられないときの対処

  答えられない場合は無理をしてあいまいな返答をしたりせず、わからないことは素直に認めて、調べて後日回答する旨を伝える

  （例）

  - That's a good question, but I'm afraid I can't answer that now. Let me check and get back to you later.

☞ 【質問シートの例】（p. 80）

  質問シートを使って、質問の練習をしよう。以下の点に留意すること。

  - 事前：事前に発表テーマについて調べておき、Step 1 から 3 に従って、質問を 1 つ用意しておく
  - オン・ザ・スポット：発表を聞きながら、質問が浮かべば、メモする

5 各種フォーマット

ハンドアウトの例

Engines

Presenter: [　　　　　　　　　　　　　　　　]

I. What Are Engines?
 1. Definition
 2. Types of Engines
 3. Combustion Engines
  (1) Internal Combustion Engines
  (2) External Combustion Engines

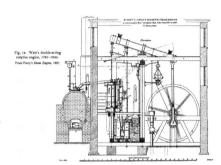

II. History of Engines
 1. Early Times
 2. Industrial Revolution
 3. Modern Times

III. Gasoline Engines and Diesel Engines
 1. Structure
 2. Function
 3. Differences

IV. Conclusion
 1. Summary
 2. Future Development of Engines

Engines

Greeting

Good afternoon and welcome to our presentation.

Purpose

Can you imagine life without engines? // Without
※ポーズ

engines, there would be no cars, no buses, no
※たたみかけるように

airplanes. There would be no electricity either,

because our power stations are run by turbines,

which are also engines. // Now, our life is

dependent on engines so much that we cannot live

without them. Today, we would like to focus on

this wonderful human invention and re-evaluate

the vital role that engines play in our daily life.

Overview

Here is an overview of our presentation. Please

take a look at the handout.

(Member Introduction)

First, Mr. x x x will define what engines are

and show us what types of engines we have today.

Second, Ms. x x x will talk about . . .

プロンプター・カードの例

I. What Are Engines?
1. Definition
   • "An engine is a machine that can convert any of various forms of energy into mechanical power or motion."—*Encyclopedia Britanica*
   • Windmills, water mills → early forms of engines
2. Types of Engines
   (1) Most engines turn heat energy into movement.
      1. Steam engines and turbines
      2. Internal combustion engines
      3. Jet engines
   (2) Steam engine
      - The first steam engine → Thomas Newcomen (1712)
      - Power source: steam pressure → fuel to boil water is burned outside the engine (= external combustion engine)

プレゼンテーション・スライドの例

Diesel Engine

1. **Structure**
   • fuel injector
   • piston and cylinder
   • crank shaft
2. **Function**
   Four Strokes
   ① air intake
   ② air compression
   ③ combustion
   ④ exhaustion

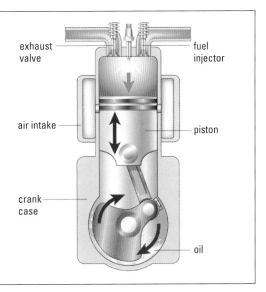

exhaust valve

fuel injector

air intake

piston

crank case

oil

Evaluation Sheet

Group Members: _____

Theme: _____

Topic	1) contents	1	2	3	4	5
	2) organization	1	2	3	4	5
Delivery	3) English	1	2	3	4	5
	4) poise	1	2	3	4	5
	5) persuasion	1	2	3	4	5

Total Score

[1) ____ + 2) ____ + 3) ____ + 4) ____ + 5) ____] × 4 =

/100

Comments

Reviewer: _____

Question Sheet

Student ID (   )   Name (   )

Prepared Question

Step 1

Step 2

Step 3

Impromptu Question

Appendix 2

Glossary

光・音・力

1. 光の反射   **reflection**
We can see things because of the **reflection** of light.
(私たちは、光の反射によって物を見ることができる)

2. 反射の法則   **law of reflection**
According to the **law of reflection**, the angle of incidence equals the angle of reflection.
(反射の法則によれば、入射角は反射角に等しい)

3. 光の屈折   **refraction**
Light bends when it enters at an angle from one transparent substance, such as air, into another substance, such as water. This bending of light is called **refraction**.
(光はある透明な物質［空気など］から、ある角度で別の物質［水など］へ入るときに折れ曲がる。このように光が折れ曲がることを屈折と呼ぶ)

4. 全反射   **total reflection** （光通信分野：total internal reflection）
Total reflection occurs when a ray of light strikes a medium boundary at an angle greater than a particular critical angle.
(全反射は、光線が媒質の境界面に対して、特定の臨界角を超えて入射した場合に起こる)

5. 焦点   **focus**
A **focus** is the point where reflected or refracted light rays converge.
(焦点とは、反射光ないし屈折光が集まる点のことである)

6. 実像   **real image**
A **real image** is formed when an object is placed outside the focal length of a convex lens.
(実像は、物体が凸レンズの焦点距離の外側に置かれたときに生ずる)

7. 虚像   **virtual image**
A **virtual image** is formed when an object is placed inside the focal length of a convex lens.
(虚像は、物体が凸レンズの焦点距離の内側に置かれたときに生ずる)

8. 振幅   **amplitude**
The **amplitude** of a sound wave is related to the loudness of the sound.
(音波の振幅は、音の大きさにかかわっている)

9. 波長   **wavelength**
The **wavelength** of a sound wave is related to the pitch of the sound.
(音波の波長は、音の高さにかかわっている)

10. 振動数   **frequency**
The **frequency** of a sound wave is the number of vibration cycles within a set period of time.
（音の周波数とは、一定時間における、音波の振動数である）

11. ニュートン   **newton**（N）
The **newton** is a unit of force. One newton (1 N) is the force of the earth's gravity on an object with a mass of about 100 g.
（ニュートン (N) は力の単位である。1 N は、およそ 100 g の物体にかかる重力の大きさに等しい）

12. 力の 3 要素 **three elements of force**
The **three elements of force** are: (1) the point of application of force, (2) the magnitude of force, (3) the direction of force.
（力の 3 要素とは、(1) 力の作用点、(2) 力の大きさ、(3) 力の向き、である）

13. 2 力のつり合いの **conditions for the equilibrium of two forces**
条件   The **conditions for the equilibrium of two forces** are: (1) The two forces are on the same line; (2) The directions of the two forces are opposite; (3) The two forces have the same magnitude.
（2 力のつり合いの条件とは、(1) 2 つの力は同一線上にある、(2) 2 つの力の向きは逆である、(3) 2 つの力の大きさは等しい、である）

14. 圧力   **pressure**
The tire needs more air **pressure**.
（そのタイヤはもっと空気圧が要る）

15. 大気圧   **atmospheric pressure**
Atmospheric pressure decreases at higher altitudes.
（高度の高いところでは、大気圧は減少する）

電気・磁気

16. 静電気   **static electricity**
Static electricity is produced by friction and it stays on the surface unless discharged.
（静電気は摩擦によって生じ、放電されない限り、表面にとどまる）

17. 電気回路   **electric circuit**
A simple **electric circuit** is made up of a battery and a resistor.
（簡単な電気回路は、電源と抵抗器から成っている）

18. 直列回路   **series circuit**
In a **series circuit**, the current through each of the resistors is the same.
（直列回路では、それぞれの抵抗器に流れる電流は等しい）

19. 並列回路   **parallel circuit**
In a **parallel circuit**, the voltage across each of the resistors is the same.
（並列回路では、それぞれの抵抗器にかかる電圧は等しい）

20. 電流   **electric current**
The SI unit of **electric current** is the ampere.
（電流の国際単位は、アンペアである）

21. 電圧   **electric voltage**
Electric voltage is defined as the electric potential difference between two points and can be regarded as a force or pressure which pushes the electric current.
（電圧は 2 点間の電位差であり、電流を押し流す力ないしは圧力ととらえることができる）

22. 電気抵抗   **electric resistance**
Electric resistance is a material's opposition to the flow of electric current, which is measured in ohms.
（電気抵抗とは、ある物質の電流の流れにくさのことを指し、単位オーム（Ω）で測られる）

23. オームの法則   **Ohm's law**
The formula of **Ohm's law** is $V = IR$, where V is voltage, I is current, R is resistance.
（オームの法則の公式は、$V = IR$ ［V は電圧、I は電流、R は抵抗］で表される）

24. 導体
（電気伝導体）   **electric conductor**
Copper is a good **electric conductor**.
（銅は、よく電気を通す物質だ）

25. 絶縁体   **insulator**
Some common **insulators** include glass, plastic, and rubber.
（代表的な絶縁体には、ガラス、プラスティック、ゴムなどがある）

26. 電力   **electric power**
Electric power is the amount of work done by an electric current per unit time and measured in watts.
（電力とは、電流が単位時間にする仕事量のことで、ワット (W) で表される）

27. ワット   **watt**（W）
One **watt** (1 W) is the power of 1 ampere of electric current when the potential difference is 1 volt.
（1 ワットは、1 ボルトの電圧をかけて 1 アンペアの電流が流れたときの電力である）

28. 電力量   **electric energy**
Electric energy is the amount of work that can be done by electricity, i.e., the amount of electric power produced or used over a specific period of time. The unit of measurement is the joule (1 J = 1 Ws). In a general household, electric energy is measured in watt-hours (Wh) or kilowatt-hours (kWh).
（電力量は、電気によってなされる仕事量、すなわち一定時間に生産または消費される電力の量のことである。単位はジュール［1 J = 1 Ws］。一般家庭の電力量は、ワット時 (Wh)、またはキロワット時 (kWh) で測られる）

29. 熱量   **heat energy**

Heat energy is measured in joules (J) or calories (cal). One joule (1 J) is the amount of heat generated by 1 W of electricity in a second. One calorie (1 cal) is the amount of heat required to raise the temperature of 1 g of water by 1°C. (1 J ≒ 0.24 cal)

（熱量はジュール (J) またはカロリー (cal) で測られる。1 ジュールは、1 ワットの電力で 1 秒間に発生する熱量であり、1 カロリーは、1 g の水の温度を 1°C 上昇させる熱量である。[1 J ≒ 0.24 cal]）

30. 磁力   **magnetic force**

A cylindrical magnet has a strong **magnetic force**.

（円筒磁石は、強い磁力を持っている）

31. 磁界（磁場）   **magnetic field**

Magnetic resonance imaging (MRI) is a diagnostic scanning system that has a strong **magnetic field**.

（MRI [磁気共鳴画像診断装置] は、強力な磁場を持った診断システムである）

32. 磁界の向き   **direction of a magnetic field**

The **direction of a magnetic field** is the direction the north end of a compass will point when placed in a magnetic field. It is represented by the lines of magnetic force.

（磁場の向きは、磁場に置かれたコンパスの磁針が N を指す向きである。磁場の向きは、磁力線で表すことができる）

33. 磁力線   **lines of magnetic force**

Lines of magnetic force come out of the north pole and go into the south pole.

（磁力線は N 極から出て、S 極に入る）

34. 右ねじの法則   **right-handed screw rule (Ampère's law)**

The **right-handed screw rule** states that if you wrap your right hand around a straight wire with your thumb pointing in the direction of the electric current, your fingers will curl in the direction of the magnetic field. It also states that if you curl your right hand around a coil (solenoid) with your fingers in the direction of the electric current, your thumb will point in the direction of the magnetic lines in the coil.

（右ねじの法則 [アンペールの法則] は、まっすぐな導線に電流が流れるとき、電流の向きに右手の親指を合わせて握るようにすると、磁界の向きは他の指が指す方向になることを意味する。また、右ねじの法則は、円筒コイルに電流が流れるとき、電流の向きに右手の指を合わせて握るようにすると、親指がコイル内の磁力線の向きを指すことを意味する）

35. 左手の法則   **Fleming's left-hand rule**

Fleming's left-hand rule states that if you hold your left hand with the thumb, index finger and middle finger all at right angles to each other so as the index finger points in the direction of the field (north to south) and the middle finger (second finger) in the direction of the current, the thumb will point in the direction of the force.

（フレミングの左手の法則は、左手の親指、人差し指、中指をたがいに直角になるようにし、人差し指を磁界の向き、中指を電流の向きに合わせると、親指の指す向きが、電流が磁界から受ける力の向きになることを意味する）

36. 電磁誘導　　**electromagnetic induction**
The principle of **electromagnetic induction** is used in most electric machines.
（電磁誘導の原理は、ほとんどの電気機械で用いられている）

37. 誘導電流　　**induced current**
The generated **induced current** was stored in a capacitor.
（発生した誘導電流は、コンデンサに蓄えられた）

38. 発電機　　**electric generator (dynamo)**
An **electric generator** is a device that can convert kinetic energy into electric energy.
（発電機は、運動エネルギーを電気エネルギーに変える装置である）

運動とエネルギー

39. 速さ　　**speed**
Speed is the distance traveled in a period of time.
（速さとは、一定時間に移動する距離のことである）

40. 平均の速さ　　**average speed**
The **average speed** of a moving object is calculated by dividing the distance traveled by the time of travel.
（移動物体の平均速度は、その物体が移動した距離を、それが要した時間で割ることによって計算される）

41. 瞬間の速さ　　**instantaneous speed**
The **instantaneous speed** is the speed at a particular instant of time, and it may change from moment to moment.
（瞬間速度とは、ある瞬間の速さであり、刻一刻と変化しうる）

42. 等速直線運動　　**linear uniform motion**
If we throw a ball in outer space, it will perform a **linear uniform motion**.
（宇宙空間でボールを投げると、それは等速直線運動をするだろう）

43. 摩擦力　　**friction**
Friction is the force that resists the motion of an object that is in contact with another object.
（摩擦力とは、接触している物体の動きに抵抗しようとする力のことである）

44. ニュートンの運動の法則　　**Newton's laws of motion**
Newton's laws of motion are: the law of inertia, the equation of motion, and the law of action and reaction.
（ニュートンの運動の法則とは、慣性の法則、運動方程式、作用反作用の法則、である）

45. 慣性の法則　　**law of inertia**
The **law of inertia** states that an object will stay at rest or move in a straight line at a constant speed unless it is acted upon by a force.
（慣性の法則とは、力が加えられなければ、物体は静止したままか、等速直線運動を続けることをいう）

46. 運動方程式　　　**equation of motion**
Newton's second law of motion is known as the **equation of motion** $F = ma$, where F is force, m is mass and a is acceleration.
（ニュートンの運動の第 2 法則は、運動方程式 $F = ma$［F は力、m は質量、a は加速度］として知られている）

47. 作用と反作用　　**action and reaction**
According to the law of **action and reaction**, when one object exerts a force on another object (action), the second object exerts an equal force in the opposite direction (reaction).
（作用と反作用の法則によれば、ある物体に力を加えると［作用］、同時にその物体から逆向きで同じ大きさの力を受ける［反作用］）

48. 仕事　　　　　　**work**
In physics, **work** means exerting a force to move an object in the direction of the force. Work can be zero if the object does not move at all no matter how much force may be applied.
（物理学において、「仕事」とは物体に力を加えてその向きに物体を動かすことを意味する。たとえ物体にいくら力を加えても、それが動かなければ仕事をしたことにはならない）

49. 仕事量　　　　　**amount of work**
The **amount of work** is measured in joules (J). One joule (1 J) is the amount of work done when 1 N of force acts through a distance of 1 m.
（仕事の大きさはジュール (J) で表される。1 ジュールは、1 ニュートンの力で 1 メートル動かしたときの仕事量を表す）

50. 仕事の原理　　　**law of work**
According to the **law of work**, the amount of work remains constant no matter how long it takes.
（仕事の原理によれば、仕事の総量は、仕事にかかった時間にかかわらず一定である）

51. 仕事率　　　　　**power**
Power is the amount of work done in a period of time, i.e., the rate at which the work is done. Power = Work / Time.
（仕事率とは、単位時間当たりの仕事量のことである）

52. エネルギー　　　**energy**
Energy is the ability to do work. Therefore, the amount of energy is measured by the amount of work that can be done by the energy.
（エネルギーとは仕事をする能力のことであり、それゆえにエネルギーの量は、そのエネルギーによってなされ得る仕事の量によって測られる）

53. 位置エネルギー　**potential energy**
Potential energy is energy stored in an object where there is a force that tends to pull the object back towards some lower energy position.

（位置エネルギーとは、物体をより低いエネルギー位置に引き戻そうとする力が存在するときに、その物体の中に貯えられているエネルギーのことである）

54. 運動エネルギー　**kinetic energy**
Kinetic energy is the energy that a moving object possesses because of its speed and mass.
（運動エネルギーとは、運動する物体が、その速度と質量のゆえに持つエネルギーのことである）

55. 力学的エネルギー　**mechanical energy**
Mechanical energy is the sum of potential energy and kinetic energy.
（力学的エネルギーとは、位置エネルギーと運動エネルギーを合わせたものである）

56. 力学的エネルギー　**law of conservation of mechanical energy**
保存の法則　The **law of conservation of mechanical energy** states that the sum of potential energy and kinetic energy remains constant.
（運動エネルギー保存の法則とは、位置エネルギーと運動エネルギーの総和が常に一定に保たれることをいう）

57. エネルギー　**law of conservation of energy**
保存の法則　The **law of conservation of energy** states that energy cannot be created nor destroyed. Thus the total amount of energy is constant even though energy may change forms.
（エネルギー保存の法則によれば、エネルギーは新たに生み出されることはなく、消滅することもない。それゆえ、エネルギーは形を変えることがあっても、その総量は一定である）

クラス用音声CD有り（非売品）

Basic English for Physics [Text Only]

理工系学生のための基礎英語：物理

2011 年 1 月 20 日　初版発行
2024 年 9 月 20 日　Text Only版第 1 刷

著　　者　　井村　誠
監 修 者　　伊東 健治
英文校閲　　Damien Healy、Matthew Caldwell
発 行 者　　松村 達生
発 行 所　　センゲージ ラーニング株式会社
　　　　　　〒 102-0073　東京都千代田区九段北 1-11-11 第 2 フナトビル 5 階
　　　　　　電話　03-3511-4392
　　　　　　FAX　03-3511-4391
　　　　　　e-mail: eltjapan@cengage.com
　　　　　　copyright © 2011 センゲージ ラーニング株式会社

装丁・組版　　㈱興陽社
本文イラスト　㈱ユニオンプラン
印刷・製本　　株式会社エデュプレス

ISBN 978-4-86312-414-1